GODWIN STUDY GUI

MECHANICS OF FLUIDS

Other titles in this series

Materials and Structures
Theory of Structures
Advanced Theory of Structures
Structural Steelwork
Design of Reinforced Concrete Elements
Specifications and Quantities
Surveying
Soil Mechanics
Mathematics for Engineers
Fluid Mechanics

GODWIN STUDY GUIDES

MECHANICS OF FLUIDS

R. H. DUGDALE

B.Sc.(Eng.), M.Sc., F.C.I.O.B., A.M.B.I.M., A.C.G.I.
Head of Department of Construction & Surveying,
Erith College of Technology

and

J. C. MORFETT

M.Sc.(Eng.), C.Eng., M.I.Mech.E.
Senior Lecturer in Fluid Mechanics,
Brighton Polytechnic

GEORGE GODWIN

London and New York

George Godwin
an imprint of:
Longman Group Limited
Longman House, Burnt Mill, Harlow
Essex CM20 2JE, England
Associated companies throughout the world

Published in the United States of America
by Longman Inc., New York

First published 1983

British Library Cataloguing in Publication Data

Dugdale, R. H.
Mechanics of fluids. – (Godwin study guides)
1. Fluid mechanics
I. Title II. Morfett, J. C.
532 QC145.2

ISBN 0–7114–5637–2

Printed in Hong Kong by
Astros Printing Ltd.

CONTENTS

GENERAL INTRODUCTION

THIS series was originally designed as an aid to students studying for technical examinations, the aim of each book being to provide a clear concise guide to the *basic principles* of the subject, reinforced by worked examples carefully selected to illustrate the text. The success of the series with students has justified the original aim, but it became apparent that qualified professional engineers in mid-career were finding the books useful.

In recognition of this need, the books in the series have been enlarged to cover a wider range of topics, whilst maintaining the concise form of presentation.

It is our belief that this increase in content should help students to see their study material in a more practical context without detracting from the value of the book as an aid to passing examinations. Equally, it is believed that the additional material will present a more complete picture to professional engineers of topics which they have not had occasion to use since completing their original studies.

A list of other books in the series is given at the front of this book. Further details may be obtained from the publishers.

M. J. Smith
General Editor

AUTHORS' PREFACE

This book is intended both for engineering students and practising engineers, and it covers the fundamental principles of fluid mechanics. These principles are illustrated by reference to a wide range of engineering applications. The text should, therefore, meet the needs of those engaged in mechanical, chemical, process or air-conditioning engineering.

The standard adopted is that of an engineering degree. Consequently the book should also be of use to those studying for the Part II examinations of the Council of Engineering Institutions or for the TEC Higher Certificate/Diploma (or the HNC/HND).

The teaching philosophy parallels that of the companion volume for civil engineers (by R.H.D.). That is to say the emphasis is placed on the basic principles (continuity, energy and momentum), and their applications are demonstrated through the substantial number of worked examples in the text. In general, the treatment is mathematical and the reader should therefore have a good pocket calculator. A programmable calculator is particularly useful where iterative methods are adopted.

Some of the examples in the text are our own, others were derived from the following sources:

CEI Part II examinations in Fluid Mechanics.

HNC, HND and Degree examinations papers.

The authors therefore acknowledge the courtesy of the CEI and the School of Engineering at Thames Polytechnic for permission to use examination questions. Thanks are also due to colleagues (Dr. J. Kearsey, Mr. A. Chadwick and Mr. M. Rees) for their constructive comments during the preparation of the book.

January 1983 *R.H.D.* *J.C.M.*

LIST OF PRINCIPAL SYMBOLS

Note: In view of the large number of variables in fluid mechanics it is inevitable that some symbols have to be used for more than one variable. The context will make usage clear in such cases.

Symbol	*Meaning*
a	acceleration, area, distance
A	area
b	width
c	speed of sound
C	Chezy coefficient
C_d	discharge coefficient
C_D	drag coefficient
C_f	friction coefficient
C_L	lift coefficient
C_p	specific heat at constant pressure
C_v	specific heat at constant volume
d	depth, diameter
D	drag force
e	internal energy of a fluid
F	effort, force
f	Darcy friction factor
F_r	Froude number
g	gravitational acceleration
h	enthalpy of a fluid
h	height or depth, head
$\bar{h}$	distance to centre of pressure
H	head
H_i	ideal head (for fluid machine)
h_f	head loss due to friction
h_L	head loss
H_m	manometric head (of fluid machine)
i	hydraulic gradient, impeller/runner velocity
I	second moment of area
k	permeability coefficient, wall roughness
K_H	head coefficient (for fluid machine)
K_P	power coefficient (for fluid machine)
K_Q	discharge coefficient (for fluid machine)
L	length
m	mass, hydraulic mean radius
$\dot{m}$	mass flow
M	metacentre
M_a	Mach number
n	Manning friction coefficient
p	pressure
P	force, power, wetted perimeter
q	heat transfer per unit mass flow, 2-dimensional flow
Q	discharge, heat transfer
r	radius
R	gas constant, drag force, resultant force
R_e	Reynolds number
s	entropy
T	temperature, time
u	velocity
u'	fluctuating (turbulent) velocity
u_*	friction (shear) velocity
$\bar{u}$	mean velocity
U	average velocity, free stream velocity, terminal velocity
v	velocity, specific volume, volume
v'	fluctuating (turbulent) velocity
v_f	flow velocity in fluid machine
V	velocity, volume
V_∞	free stream velocity
w	whirl velocity, work per unit mass flow
W	work done, weight, load
W_e	Weber number
x	distance (usually horizontal)
y	distance (usually vertical)
$\bar{y}$	vertical distance to centroid
z	height

Greek symbols	*Meaning*
α	angle
Γ	circulation
γ	adiabatic index
δ	boundary layer thickness
δ_d	displacement thickness
δ_e	energy thickness
δ_m	momentum thickness
η	efficiency
θ	angle
μ	absolute viscosity, vortex strength
ν	kinematic viscosity
ξ	vorticity
ρ	density
σ	surface tension
σ_T	Thoma cavitation parameter
τ	shear stress
τ_0	boundary shear stress
ϕ	angle, pressure ratio, velocity potential
ψ	stream function

CHAPTER 1

HYDROSTATICS

INTRODUCTION

System of units

Throughout this book the International System of Units is adopted. The full title "Système International d'Unité" is abbreviated to "SI" units in all languages.

In each problem, regardless of what units the data are given in, length will be converted into metres, mass into kilogrammes and time into seconds for the purpose of calculation. The proper units of force are then *Newtons.*

Mass and force are related by Newton's second law of motion:

$$P = ma$$

where P is the force in Newtons (N) which, acting on a mass m kilogrammes (kg), produces an acceleration of a metres per second per second (m/s^2). The equation can be rewritten as:

$$1\ (\text{N}) = 1\ (\text{kg}) \times 1\ (\text{m/s}^2)$$

On no account should force be expressed in kg (or kgf). Throughout this book the *local gravitational acceleration* g will be taken as 9·81 m/s^2. Since a freely falling body of mass 1 kg will accelerate at g m/s^2, the force causing the acceleration (*i.e.* its weight) will be $P = 1 \times 9{\cdot}81$ N. In other words a *mass* of 1 kg *weighs* 9·81 N.

Terms used in hydrostatics

Hydrostatics is the study of liquids at rest. Many of the terms defined in this section, however, apply equally in *hydrodynamics*, the study of liquids in motion.

Density (ρ) (rho) is mass per unit volume and, in keeping with the above remarks, should always be expressed in kg/m^3 for the purpose of calculations. For water,

$$\rho = 1000\ \text{kg/m}^3, \quad \textit{i.e.}\ 1{\cdot}0\ \text{Mg/m}^3$$

where 1 megagramme (Mg) = 1000 kg

Specific weight (ρg) is the weight per unit volume. Many writers use the symbol w, but here the double symbol ρg is preferred, as it emphasises that specific weight has a different numerical value from density. It is also important to remember that specific weight is a direct function of the local gravitational acceleration g. For water (strictly, at 4° C) $\rho g = 9{\cdot}81$ kN/m^3.

The *relative density* of a liquid is the density of the liquid expressed as a ratio of the density of water. Thus if a liquid has a relative density of 0·8 its density is 0·8 Mg/m^3 and its specific weight is $0{\cdot}8 \times 9{\cdot}807$, *i.e.* 7·846 kN/m^3.

Pressure (p) is force per unit area and is measured in N/m^2. Fluid pressure exhibits the following properties:

(*a*) At any point within a fluid, the pressure is the same in all directions; and

(*b*) The resultant pressure on the walls of the container of the fluid always acts normal to the surface.

These properties are made use of in, for example, a hydraulic jack in which the effort, F, is applied to a small piston of area a in a cylinder containing oil. This cylinder is connected by a pipe to a large cylinder of area A, the whole system being filled with oil. The rise in pressure in the large cylinder is the same as the rise in pressure in the small one, namely, F/a. Consequently, a piston in the large cylinder would support a load $W = A.F/a$, giving a mechanical advantage of $W/F = A/a$, if the jack were 100% efficient (*see* Fig.1).

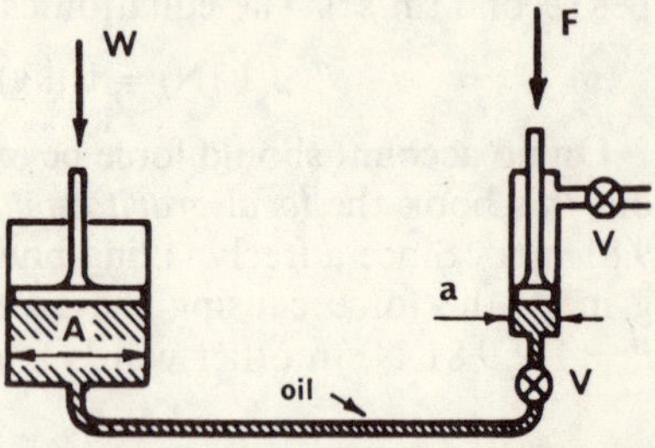

FIG. 1. The principle of the hydraulic jack. The valves which regulate the supply and flow of oil are labelled V

Absolute pressure. At any point within a perfect vacuum the absolute pressure is zero. The absolute pressure at a point in a fluid is the pressure reckoned from this zero, and will always be positive. In fluid mechanics there are two common ways of expressing pressure: in N/m^2 and in m *head*. Thus, under normal conditions atmospheric pressure is approximately 101·337 kN/m^2 absolute, or 10·33 m head of water.

To convert from m head of water into N/m^2 it is convenient to imagine the column of water as having a cross-sectional area of 1 m^2. Then the pressure at the foot of a 10·33 m column is $10{\cdot}33 \times 9{\cdot}81 \times 10^3$ N/m^2 = 101·337 kN/m^2.

It should be noticed that a 10·33 m depth of water gives the same liquid pressure, whatever the shape of the vessel containing the water, by virtue of the property that fluid pressure at a point is the same in all directions.

Gauge pressure. Usually, in fluid mechanics, pressures are reckoned from atmospheric pressure. In other words, atmospheric pressure is taken as zero, giving a scale of pressures which differs from the absolute scale. Pressures on this scale are described as *gauge pressures*, and are related to absolute pressures by the following equation:

$$\text{Gauge pressure} = \text{Absolute pressure} - \text{Atmospheric pressure}$$

MEASUREMENT OF PRESSURE

Of the large number of different kinds of pressure-measuring devices, three types of manometer are shown in Fig. 2 and one type of differential pressure gauge is the subject of Example 2.

In Fig. 2 (*a*) liquid of density ρ is flowing along the pipe. If the piezometer is tapped into the pipe the liquid will rise to height h above the centre-line. The pressure at the centre-line is then said to be h m of the liquid or, alternatively, $\rho g h$ N/m^2 gauge.

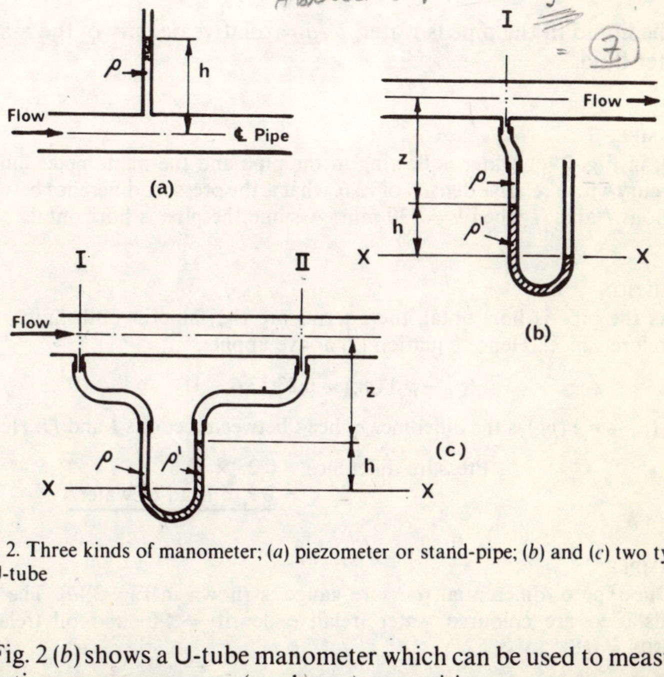

FIG. 2. Three kinds of manometer; (*a*) piezometer or stand-pipe; (*b*) and (*c*) two types of U-tube

Fig. 2 (*b*) shows a U-tube manometer which can be used to measure negative gauge pressures (as shown) or positive gauge pressures of greater magnitude than can conveniently be measured by a piezometer. The density of the liquid flowing along the pipe is ρ and the density of the liquid filling the bend of the U-tube is ρ'. To find the pressure at the centre-line of the pipe at section *I*, note that the pressure in the two limbs of the manometer at section *X X* must be equal, for otherwise the liquid in the bend would move, being acted on by different forces. But the pressure on the right-hand limb is atmospheric, while the pressure on the left-hand limb at *XX* is p_1, the pressure at the centre-line of the pipe, plus $\rho gz + \rho' gh$. Hence,

$$p_1 + \rho gz + \rho' gh = \text{atmospheric pressure}$$

$$\therefore p_1 = -\rho gz - \rho' gh, \text{ gauge pressure}$$

Fig. 2 (*c*) shows an adaptation of the U-tube for comparing the pressures at sections *I* and *II* of the pipe. By equating the pressures at *XX* as above, the difference between the pressure at section *I*, p_1, and at section *II*, p_2 is

$$p_1 - p_2 = \rho' gh - \rho gh$$

provided that the pipe is horizontal. This can, alternatively, be written

$$(p_1 - p_2)/(\rho g) = h(\rho'/\rho - 1) \qquad (1)$$

If the liquid in the pipe is water, ρ'/ρ = relative density of the manometer fluid.

EXAMPLE 1

If, in Fig. 2 (*c*), water is flowing in the pipe and the manometer fluid is mercury with a relative density of 13·6, what is the pressure difference between sections *I* and *II* when $h = 230$ mm? Assume the pipe is horizontal.

SOLUTION

As the pipe is horizontal, the *z*-terms are the same in both limbs, and therefore cancel. Hence equation (1) above applies:

$$(p_1 - p_2)/(\rho g) = 0{\cdot}23(13{\cdot}6 - 1)$$

But $(p_1 - p_2)/(pg)$ is the difference in head between sections *I* and *II*. Hence:

$$\text{Pressure difference} = 0{\cdot}23 \times 12{\cdot}6$$
$$= \underline{\underline{2{\cdot}9 \text{ m head of water.}}}$$

EXAMPLE 2

One type of differential pressure gauge is shown in Fig. 3 (*a*). The two fluids used are coloured water (relative density = 1·0) and oil (relative density = 0·8).

(*a*) Work from first principles to determine:

(*i*) the dimension h when $p_1 = p_2$;
(*ii*) the pressure difference $p_2 - p_1$ that would cause the water level "*A*" to rise 1·5 cm.

(*b*) What is the particular function of such a manometer?

(*c*) Why must the oil surface "*B*" lie in the widened part of the right-hand limb when readings are taken?

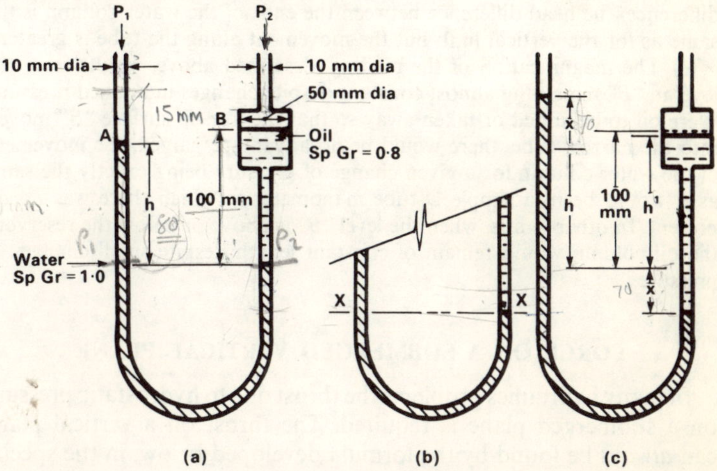

FIG. 3. *See* Example 2

SOLUTION

(*a*) With reference to Fig. 3 (*b*), the key to this kind of problem is always to equate the forces acting above *X–X* which, here, passes the oil–water interface.

(*i*) Since the c.s.a. of each branch of the manometer is the same,

$$p_1 + \rho g h = p_2 + \rho' g \times 0{\cdot}1 \qquad \text{(working in metres)}$$

Since $p_1 = p_2$

$$h = \frac{\rho'}{\rho} \times 0{\cdot}1 = 0{\cdot}08 \text{ m} = \underline{\underline{80 \text{ mm}}}$$

(*ii*) When level "*A*" rises by x the level in "*B*" falls by $x \times \left(\frac{1}{5}\right)^2$

i.e. by $\dfrac{x}{25} = \dfrac{15}{25} = 0{\cdot}6$ mm

$$\therefore \qquad p_1 + \frac{(80 + 2 \times 15)}{1000}\rho g = p_2 + \frac{(100 + 15 - 0{\cdot}6)}{1000}\rho' g$$

$$\therefore \qquad p_2 - p_1 = \frac{\rho g(110 - 114{\cdot}4 \times 0{\cdot}8)}{1000} = \underline{\underline{181{\cdot}3 \text{ N}}}$$

(*b*) The function of differential manometers such as this is to magnify the reading. Had a simple U-tube manometer containing water been used instead, the same pressure difference would have caused a movement between the ends of the water column of $181{\cdot}3 \times 1000 \div 9810$ mm, *i.e.* 18·5 mm compared with the movement in the differential manometer of nearly 30 mm.

The movement can be magnified further by making the left-hand limb of the manometer inclined to the horizontal. Then, for a given applied pressure difference, the head difference between the ends of the water column is the same as for the vertical limb but the movement along the tube is greater.

(*c*) The magnification of the reading, described above, results from the level in "*B*" remaining almost constant despite changes in applied pressure. Were oil to be added or taken away so that the free oil surface "*B*" moved into the narrow tube, there would be no advantage gained, the movement of the water column for a given change of pressure being exactly the same as it would be in a simple U-tube manometer in which there was no oil present. In other words, when the level "*B*" is above or below the reservoir, the oil column would remain of constant length despite small changes of pressure.

FORCE ON A SUBMERGED VERTICAL PLANE

In many hydraulics problems the thrust due to hydrostatic pressure on a submerged plane is required. The thrust on a vertical plane can *always* be found by the formula developed below; in the special case of a rectangular plane, however, diagrams showing the distribution of hydrostatic pressure on the plane can be drawn and the resulting thrust calculated from the area of these pressure diagrams.

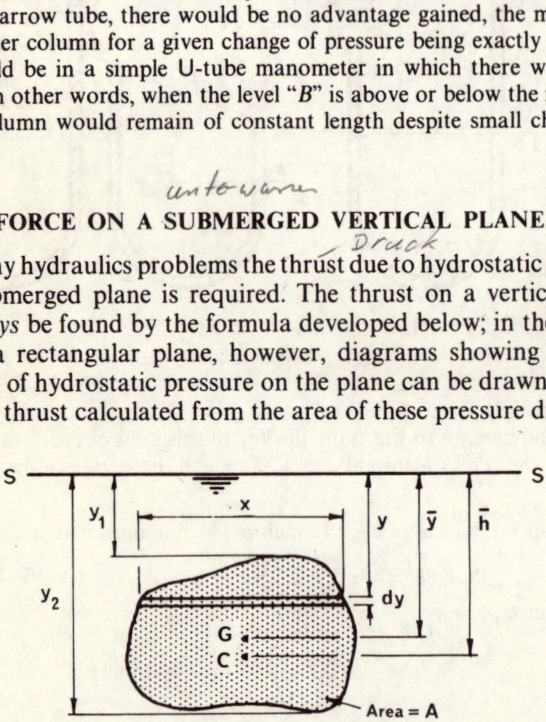

FIG. 4. An irregular-shaped submerged vertical plane

The hydrostatic force on one side of the plane shown in Fig. 4, due to liquid of density ρ, can be found by calculus:

$$\text{Pressure at depth } y = \rho g y,$$

$$\therefore \text{ Force on the element of width } dy = \text{Pressure} \times \text{Area} = \rho g y x . dy$$

Hence, over the whole area, A,

$$\text{Total force} = \rho g \int_{y_1}^{y_2} yx.\mathrm{d}y$$

$$= \rho g . A\bar{y}$$

because the first moment of area of A about the surface SS is defined as $\int_{y_1}^{y_2} yx.dy$, and is also equal to $A\bar{y}$, where $\bar{y}$ is the depth of the centroid, G, of the area. Also, $\rho g\bar{y}$ is the pressure at G. Hence,

$$\text{Total force} = \text{Area} \times \text{Pressure at } G$$

The *centre of pressure*, C, is the point at which the total force may be considered to act, and, in general, does not coincide with the centroid, G. The depth, $\bar{h}$, of C is found by taking moments about SS. Thus

$$\text{Moment of total force about } SS = \rho g A\bar{y} \times \bar{h} \text{ and}$$

Sum of the moments due to the forces on elements of width dy

$$= \rho g \int_{y_1}^{y_2} yxy.\mathrm{d}y$$

These two expressions represent the same thing. Hence,

$$A\bar{y} \times \bar{h} = \int_{y_1}^{y_2} y^2 x.\mathrm{d}y$$

Table I lists the second moment of area for the more common geometric shapes.

But $A\bar{y}$ is the first moment of area about SS *and* $\int_{y_1}^{y_2} y^2x.\mathrm{d}y$ is the second moment of the area A about SS, whence

$$\bar{h} = \frac{\text{Second moment of area } A \text{ about } SS}{\text{First moment of area } A \text{ about } SS}$$

If I_{GG} denotes the second moment of area A about an axis through G and parallel to SS, then, by the parallel-axis theorem,

$$\bar{h} = (I_{GG} + A\bar{y}^2)/(A\bar{y}) = \bar{y} + I_{GG}/(A\bar{y})$$

from which,

$$GC = \bar{h} - \bar{y} = I_{GG}/(A\bar{y})$$

TABLE I. Second moments of area.

Shape and axis	Second moment of area
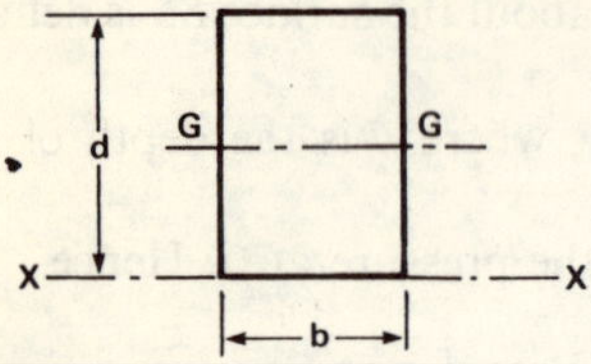	$I_{GG} = \frac{bd^3}{12}$ $I_{XX} = \frac{bd^3}{3}$
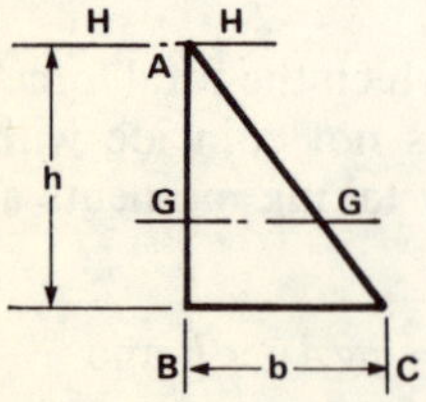	$I_{BC} = \frac{bh^3}{12}$ $I_{HH} = \frac{bh^3}{4}$ $I_{GG} = \frac{bh^3}{36}$
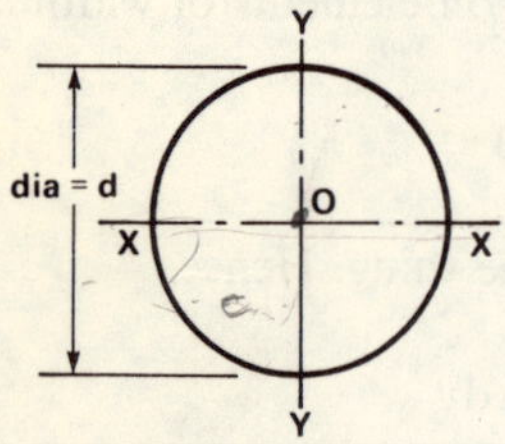	$I_{XX} = I_{YY} = \frac{\pi d^4}{64}$ $I_{polar} = \frac{\pi d^4}{32}$
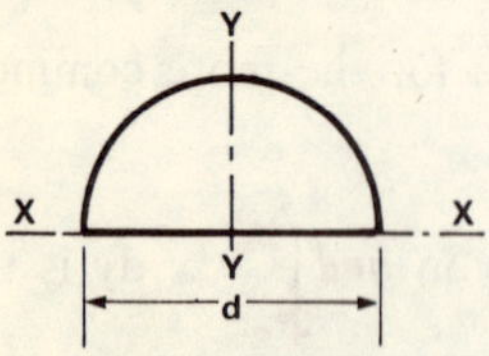	$I_{YY} = \frac{1}{2} \times \frac{\pi d^4}{64} = \frac{\pi d^4}{128}$ $I_{XX} = \frac{d^4}{128}$
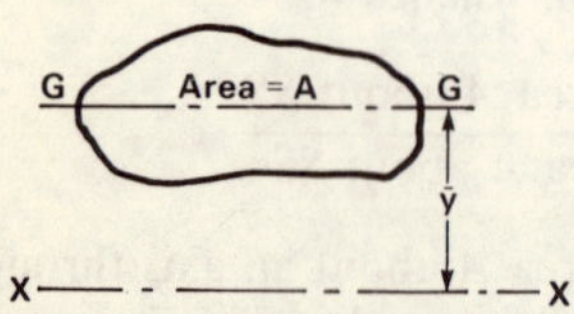	Parallel axis theorem: $I_{XX} = I_{GG} + A\bar{y}^2$ where GG is an axis through the centroid parallel to XX.

Alternatively, as

$$I_{GG} = Ak_G^{\ 2}$$

where k_G = the radius of the gyration of the plane about G, then

$$GC = k_G^{\ 2}/\bar{y}$$

i.e. C is always below G by an amount $k_G^2/\bar{y}$, which is the same as $I_{GG}/(A\bar{y})$.

In problems, $\bar{y}$ is usually known exactly. Consequently the most accurate, and often the most convenient, way of calculating $\bar{h}$ is to calculate GC by the formula and add it to $\bar{y}$.

EXAMPLE 3

Deduce expressions for the force and depth of centre of pressure on a vertical plane surface subject to water on one side.

A culvert of 2·5 m diameter circular cross-section is closed by a vertical disc of the same diameter which is pivoted about a horizontal diameter. Determine the net turning moment about the axis if the head of water upstream of the disc is 12 m and that downstream is 6 m above the axis.

SOLUTION

In Fig. 5, G is the centre of the disc and, therefore, its centroid; P_1 is the thrust due to the water upstream and acts at C. Downstream the thrust is P_2, and the centre of pressure is B.

For a circle turning about a diameter, $I_{GG} = \pi d^4/64$.
Since $A = \pi d^2/4$,

$$I_{GG}/(A\bar{y}) = (\pi d^4/64)/(\pi \bar{y} d^2/4) = d^2/(16\bar{y}) = 0{\cdot}391/\bar{y}$$

Upstream, $\bar{y} = 12$ m and, therefore, $GC = 0{\cdot}391/12 = 0{\cdot}0326$ m, and downstream, $\bar{y} = 6$ m and, therefore, $GB = 0{\cdot}0652$ m.

The forces on the disc are:

$$P_1 = A_1 \times 12 \times 10^3 \times 9{\cdot}81 \text{ and } P_2 = A_1 \times 6 \times 10^3 \times 9{\cdot}81$$

where A_1 is the area of the plate and need not be evaluated.

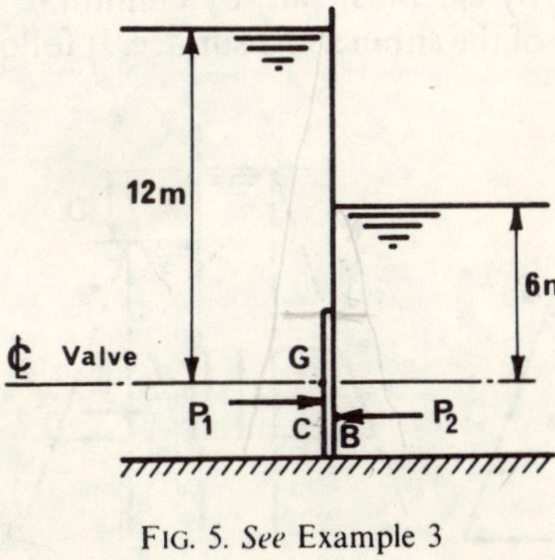

FIG. 5. *See* Example 3

Taking moments about G, the net turning moment on the disc is:

$$A_1 \times 12 \times 10^3 \times 9{\cdot}81 \times 0{\cdot}0326 - A_1 \times 6 \times 10^3 \times 9{\cdot}81 \times 0{\cdot}0652 = \underline{\underline{0}}$$

Example 3 is a particular illustration of a general result: that the turning moment on a disc valve, under the following conditions, will be zero:

(*a*) The valve must have the same wetted area both sides.
(*b*) It must be pivoted about a horizontal diameter.
(*c*) Both sides must be submerged.

Apart from these considerations, the depth of water either side of the valve is immaterial, for on the upstream side of the valve, for example, the force on the disc is proportional to y, whereas the distance GA is proportional to $1/y$, so that the product of these two quantities is independent of y—*i.e.* it is constant. It is left to the reader, as an exercise well worth while, to write out the formal proof.

The thrust due to hydrostatic pressure on *rectangular* surfaces can sometimes be found most simply by drawing pressure diagrams and basing the calculation on them (*see* Example 4).

Referring to Fig. 6 (*a*) shows a plane, vertical surface retaining water to a depth d. Triangle XYZ is the pressure diagram, and is so drawn that $Y'Z'$ represents the pressure at depth z to scale. In other words, $Y'Z'$ represents the quantity ρgz to scale. z can vary from 0 to d; to each value of z there corresponds a line, parallel to $Y'Z'$, whose length is proportional to z. Consequently the pressure diagram *is* a triangle. Moreover, YZ, the base of the triangle, scales ρgd. The total force on the wall, P, is the sum of all the pressures such as that represented by $Y'Z'$ and is therefore the area of the triangle, *i.e.*

$$P = \tfrac{1}{2}\rho g d^2$$

Furthermore, P acts through the centroid of triangle XYZ, a fact easily established by calculus. But, by definition, P also acts at the centre of pressure of the submerged surface. It follows that the centre

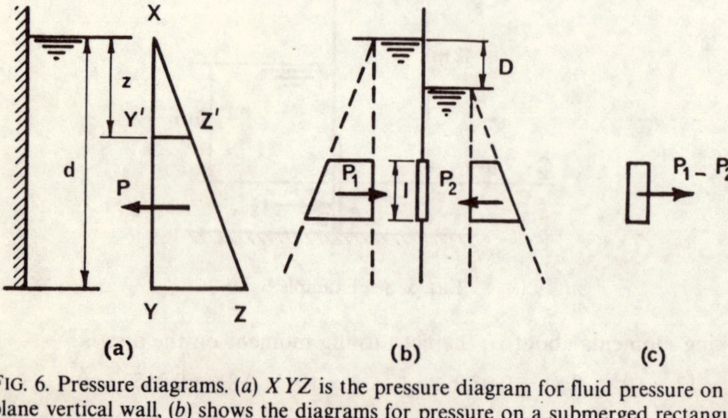

FIG. 6. Pressure diagrams. (*a*) XYZ is the pressure diagram for fluid pressure on the plane vertical wall, (*b*) shows the diagrams for pressure on a submerged rectangular valve, (*c*) is the resulting pressure diagram for (*b*)

of pressure is at the same level as the centroid of the pressure diagram, in this case at a depth of $\frac{2}{3}d$.

Fig. 6 (*b*) shows the hydrostatic pressure distribution on a rectangular valve which is submerged on both sides. The pressure diagrams are both trapezoidal; their difference, shown in Fig. 6 (*c*), is a rectangle of width ρgD because the difference between the upstream and downstream pressures at every point on the valve is ρgD. Hence the resulting thrust on the valve is

$$P_1 - P_2 = \rho gDl, \text{ per unit width of valve}$$

and acts on the horizontal centre-line of the valve.

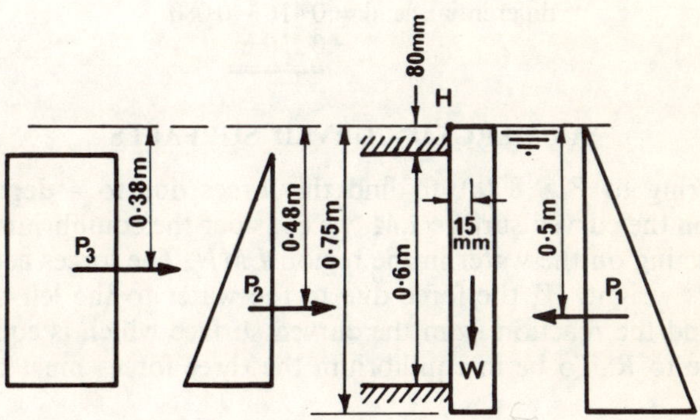

FIG. 7. Configuration, pressure diagrams and forces acting on the valve (*see* Example 4)

EXAMPLE 4

The fully submerged sea outfall of a 0·6-m-square sewer is provided with a vertical top-hung cast-steel flap valve, 0·75 m square, with the hinged edge 80 mm above the top of the sewer.

The vertical line through the centre of gravity of the flap is 15 mm outwards from the centre-line of the hinge. The weight of the flap in air is 2·5 kN; the relative density of the cast-steel is 7·75; and the density of sewage may be considered as equal to that of sea water, *i.e.* 1·028 Mg/m^3.

Calculate the differential head required just to open the flap when the surface of the sea is at the level of the hinge.

SOLUTION

Effective weight of flap, allowing for buoyancy, is:

$$W = 2{\cdot}5 \times 10^3\left(1 - \frac{1{\cdot}028}{7{\cdot}75 \times 1{\cdot}000}\right) \text{N}$$

$$= 2170 \text{ N} \qquad \text{(to 3 significant figures)}$$

Let x m = head of sewage above top of sewer. Then, referring to Fig. 7,

$$P_1 = (0{\cdot}75)^2 \times \tfrac{1}{2} \times 0{\cdot}75 \times 1028 \times 9{\cdot}81 = 2130 \text{ N}$$
$$P_2 = (0{\cdot}6)^2 \times \tfrac{1}{2} \times 0{\cdot}6 \times 1028 \times 9{\cdot}81 = 1090 \text{ N}$$
$$P_3 = (0{\cdot}6)^2 \times 1028 \times 9{\cdot}81 \times x = 3630x \text{ N}$$

Since P_1, P_2, P_3 act at the centroids of their respective diagrams,

depth of P_1 below the hinge $= \frac{2}{3} \times 0{\cdot}75 = 0{\cdot}50$ m
depth of P_2 below the hinge $= \frac{2}{3} \times 0{\cdot}6 + 0{\cdot}08 = 0{\cdot}48$ m
and depth of P_3 below the hinge $= \frac{1}{2} \times 0{\cdot}6 + 0{\cdot}08 = 0{\cdot}38$ m

Hence, taking moments about H,

$$(0{\cdot}500 \times 2130) + (0{\cdot}015 \times 2170) = (1090 \times 0{\cdot}48) + (3630x \times 0{\cdot}38)$$

from which,

$$x = 0{\cdot}416 \text{ m}$$
$$\therefore \text{ differential head} = 0{\cdot}416 - 0{\cdot}080$$
$$= 0{\cdot}336 \text{ m}$$

SUBMERGED CURVED SURFACES

Referring to Fig. 8 (*a*), to find the forces due to a depth d of water, on the curved surface LM, we consider the equilibrium of the forces acting on the water in the region LMN. The forces acting on it are its weight, W, the force due to the water to the left of NM, *i.e.* P, and the reaction from the curved surface which is equal and opposite to R. To be in equilibrium the three forces must be concurrent.

If ρ is the density of the water, and if P, W and R are forces per m run of the wall, LM,

$P = \frac{1}{2}\rho g d^2$, and acts at a height $d/3$ above M
$W = \rho g \times$ Area LMN, and acts through G, the centroid of this area, and
R is the resultant of W and P

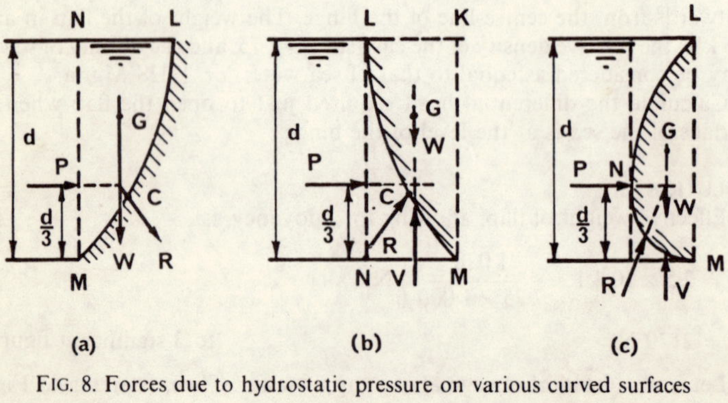

FIG. 8. Forces due to hydrostatic pressure on various curved surfaces

It will be shown that if LM is a circular arc the problem is simplified. Otherwise, before the position, magnitude and direction of R can be found it is necessary to determine where G is, so that the point of intersection of P and W can be determined.

In Fig. 8 (*b*) the forces acting on the curved surface LM are P horizontally and V vertically. In this case, V acts upwards, tending to lift LM. The forces acting on the water in the region LMN are in equilibrium (otherwise it would move). Now suppose the region KLM was filled with water, supported by a wall at KM, and then the surface LM was removed. The water in LMN would be unaffected, and we conclude, therefore, that V is equal and opposite to W, the weight of water required to fill LKM. The line of action of V must also pass through G, the centroid of LKM. Hence the position, magnitude and direction of V are known and, since the position, magnitude and drection of P are the same as in (*a*), the position, magnitude and direction of R, their resultant, can again be determined.

Fig. 8 (*c*) amounts to a combination of the two cases dealt with in (*a*) and (*b*). The horizontal component of force acting on the surface LNM is P and has the same value as in the previous two cases. The vertical component is V and, by Archimedes' principle, is equal and opposite to the weight of water displaced by the area LMN, *i.e.* W, and must line up with W, which passes through the centroid, G, of the area LNM. Again R is the resultant of V and P.

In each of the three cases discussed, R meets the surface at C, this point being defined as the *centre of pressure*. It should be noted that, although pressure always acts normal to a surface, the resultant R will not usually be normal to the surface except when the surface is circular in cross-section. For any surface, R can be thought of as the resultant of a large number of small forces which all act normal to the surface and which increase in magnitude with depth; in other words, each small force represents the pressure at the point times a small area over which it may be considered constant. When the surface is circular in section each of these forces acts through the centre of the circle, and the sum of their moments about the centre is therefore zero. Hence their resultant, R, must have zero moment about the centre, and must therefore pass through the centre.

Hence, to find the resultant force on a circular surface, it is necessary to calculate P and V (or W) in magnitude only, for these values are sufficient to determine the magnitude and direction of R. The position of R is then determined by the property that it passes through the centre of the circle.

EXAMPLE 5

What is meant by a *centre of pressure*?

How can this centre be determined for a plane submerged surface?

A steel cylinder, 2·7 m in diameter and 4·9 m in length, lies across an open rectangular channel which is 4·9 m wide and contains water. If the water depths on the upstream and downstream sides of the cylinder are respectively 2·70 m and 1·35 m, determine the magnitude, direction and location of the resulting force on the cylinder caused by the external water pressure.

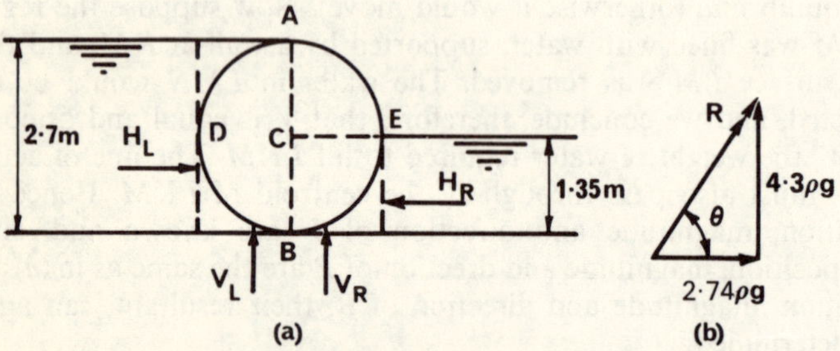

FIG. 9. *See* Example 5

SOLUTION

Initially, consider the forces acting on 1 m length of cylinder, and call these H_L, V_L, H_R and V_R (*see* Fig. 9 (*a*)). Then if *C* is the centre of the circle,

$$V_L = \text{weight of water that would fill the area } ACBD$$

and $$V_R = \text{weight of water that would fill the area } CBE$$

But these areas are, respectively, half and a quarter of the circle of diameter 2·7 m. Hence,

$$V = V_L + V_R = \tfrac{3}{4} \times (2{\cdot}7)^2\pi/4 \times \rho g = 4{\cdot}3\rho g$$

The horizontal thrusts are

$$H_L = \tfrac{1}{2} \times (2{\cdot}7)^2 \times \rho g = 3{\cdot}65\rho g$$

and $$H_R = \tfrac{1}{2} \times (1{\cdot}35)^2 \times \rho g = 0{\cdot}911\rho g$$

$$\therefore\ H = H_L - H_R = 2{\cdot}74\rho g$$

Hence, $$R = \sqrt{H^2 + V^2} = \rho g\sqrt{2{\cdot}74^2 + 4{\cdot}3^2} = 5{\cdot}1\rho g$$

and, in Fig. 9 (*b*), $\tan\theta = 4{\cdot}3/2{\cdot}74$ $\therefore\ \theta = 57{\cdot}5°$

and total thrust $= 4{\cdot}9R = 4{\cdot}9 \times 5{\cdot}1 \times 10^3 \times 9{\cdot}81 = 245$ kN.

Also, the location of *R* is such that it passes through *C*.

LOCK GATES

A standard application of hydrostatic pressure on a submerged *plane* surface occurs in the problem of designing lock gates. The purpose of lock gates is to maintain water at different levels on either side of the gates, the hydrostatic force on the gates being a maximum when the difference in level is greatest.

Fig. 10 (*a*) is a plan view in which the gates are shown to be symmetrical about the centre-line of the lock, the only case considered. The different water levels on either side of the gates are shown in the section through a gate, Fig. 10 (*b*); in this view, P_1 and P_2 are the hydrostatic forces on the gate due to the water upstream and downstream of the gate, respectively. P is the resultant of P_1 and P_2, and acts at the centre of pressure, C. It may be calculated by drawing the pressure triangles for the water on both sides of the gate, and applying the principles established earlier in the chapter.

In plan, P acts at the centre of the gate. If, moreover, the gate is not to move, P and the other forces acting on it must be in equilibrium. The other forces acting on the gate are shown in Fig. 10 (*a*) and are:

(*a*) T, *the reaction from the other gate.* Since the action on and the reaction from the other gate must be equal and opposite, and because the gates are symmetrical about the centre-line of the lock, T must be perpendicular to the centre-line of the lock.

(*b*) R, *the reaction from the hinges.* Usually there are two hinges situated near the top and bottom of each gate, respectively, and R will be the resultant of the reactions at these two hinges.

Since P, R and T are in equilibrium, they can be drawn as the sides of a triangle of forces, Fig. 10 (*c*). Furthermore, to be in equilibrium, the three forces must be concurrent in plan and, since C is the mid-point of LM in Fig. 10 (*a*), it is easy to reason that both T and R make an angle α with the gate. Fig. 10 (*c*) is therefore an

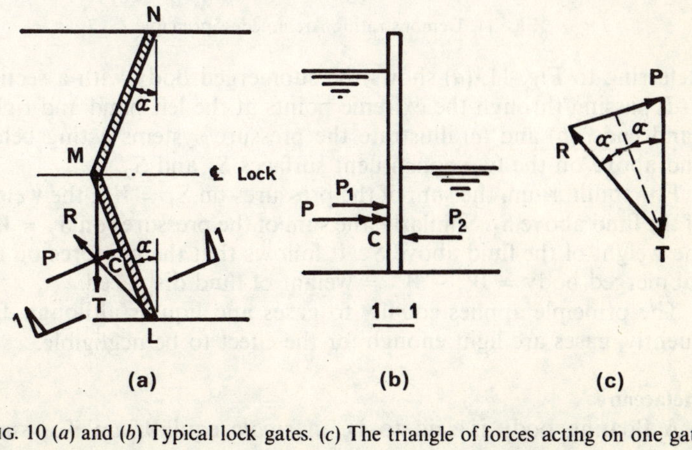

FIG. 10 (*a*) and (*b*) Typical lock gates. (*c*) The triangle of forces acting on one gate

isosceles triangle in which $T = R$. It follows from this triangle that

$$T = R = \tfrac{1}{2}P/\sin\alpha$$

Since P has already been calculated, T and R can also be found.

Again, since P, R and T are in equilibrium, they must be concurrent not only in plan but also in *space*, and consequently all three must act at the same height, namely the height of C. In particular, R acts at the height of C, and consequently, knowing the heights of the two hinges, the reaction at either can be found by taking moments about the point of action of the other.

The reader can apply the techniques described above in question 5 at the end of this chapter.

FLOATING BODIES

Archimedes' principle

That floating is a function of pressure can be demonstrated by the proof of this well-known principle which states that:

> a body which is wholly or partly submerged in a fluid suffers an apparent loss in weight equal to the weight of the fluid displaced.

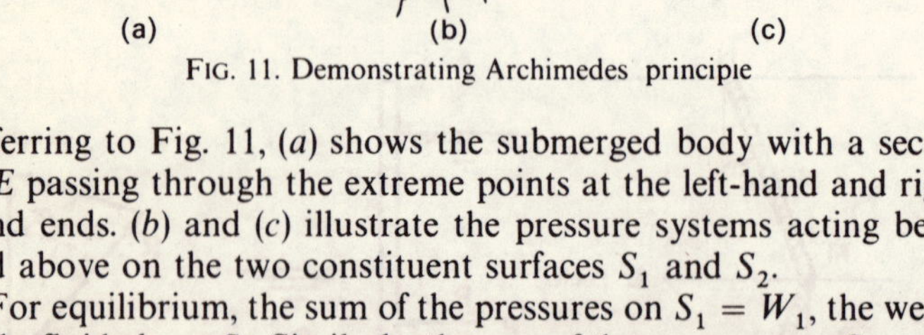

FIG. 11. Demonstrating Archimedes principle

Referring to Fig. 11, (*a*) shows the submerged body with a section E–E passing through the extreme points at the left-hand and right-hand ends. (*b*) and (*c*) illustrate the pressure systems acting below and above on the two constituent surfaces S_1 and S_2.

For equilibrium, the sum of the pressures on $S_1 = W_1$, the weight of the fluid above S_1. Similarly, the sum of the pressures on $S_2 = W_2$, the weight of the fluid above S_2. It follows that the net force on the submerged body $= W_1 - W_2 =$ weight of fluid displaced.

The principle applies equally to gases and liquids although, frequently, gases are light enough for the effect to be negligible.

Metacentre

A floating body is said to be in *stable* equilibrium if a small disturbance produces a couple which tends to restore the body to its original position. If, however, the disturbing force produces a couple which tends to increase the displacement further, the body is said to be in *unstable* equilibrium. If the disturbing force produces no couple (as in the case of a floating ball) the body is said to be in *neutral* equilibrium.

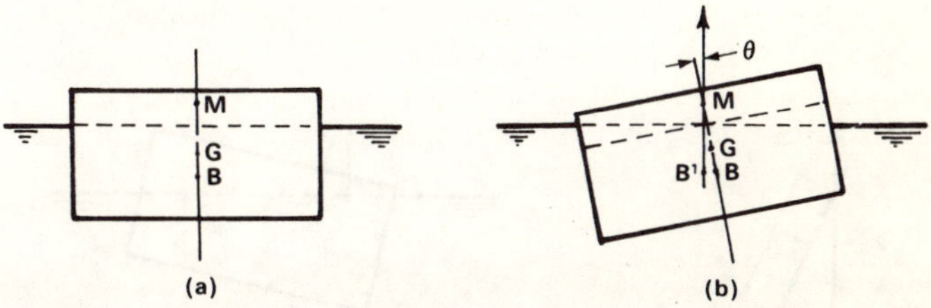

FIG. 12. Indicating where the metacentre lies

The upthrust on a submerged body acts at the *centre of buoyancy* B, which is at the centroid of the volume of liquid displaced (*see* Fig. 12).

If the centre of gravity of the body, G, is below B the body will be in stable equilibrium. If, however, G is above B the equilibrium of the body may be stable or unstable depending on shape.

With reference to Fig. 12 (*b*), the condition for stable equilibrium is that the *metacentre M* is above G, the metacentre being defined as follows. When the body is tilted through a small angle θ the centre of buoyancy moves from B to B', where B' is now the centroid of the immersed trapezium in Fig. 12 (*b*), the upthrust acts vertically through B' and intercepts the line BG produced in M thus defining the position of the metacentre. MG is called the *metacentric height*.

In Fig. 12 (*b*) the weight of the floating body is W and acts vertically downwards through G; the upthrust is also W and acts vertically upwards through M and the two together form a restoring couple tending to right the floating body.

Determination of metacentric height

An expression for MB can be obtained by considering unit length of the floating body, a ship say, and equating expressions for the shift of buoyancy which occurs when the centre-line is given a small angular displacement θ.

Let v = volume of displaced fluid per unit length of ship.

Then weight per unit length of fluid displaced = $\rho g v$.

From Fig. 13 (*a*), the couple due to this shift of buoyancy (obtained by taking moments about B) $= \rho gv.r\theta$, where $r = MB$ and θ is small.

Also, from Fig. 13 (b), taking moments about B:

$$\text{couple} = f \times \frac{2}{3}b = \rho g \times \tfrac{1}{2} \times \frac{b}{2}\theta \times \frac{b}{2} \times \frac{2}{3}b$$

$$= \rho g \times \frac{b^3\theta}{12}$$

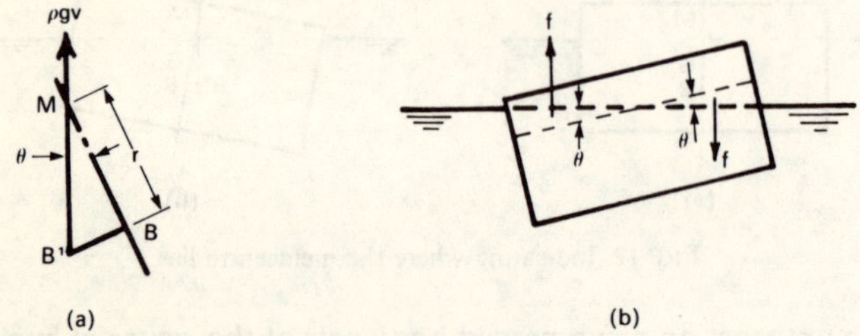

FIG. 13. Determination of metacentric height

but $\frac{b^3}{12} = i$, the second moment of area of the unit length of ship.

Hence, equating the two expressions for the couple:

$$\rho gvr\theta = \rho g\theta i$$

i.e. $$r = \frac{i}{v}$$

For the full length of the ship i has to be replaced by I, the second moment of area of the water line section, and v by V, the total volume displaced. Hence:

$$\underline{\underline{MB = r = \frac{I}{V}}}$$

The metacentric height is then calculated from $MG = MB - GB$.

EXAMPLE 6

A steel pontoon has the dimensions shown in Fig. 14. Its mass is 90 tonnes and it is one of several which support a temporary bridge which crosses a river. The centre of gravity of the pontoon is at the geometrical centre of the plan and 1·5 m above the base. The bridge trusses impose a total maximum load of 2000 kN on the centre-line of the pontoon deck.

Determine the metacentric height of the pontoon when it is loaded as shown in Fig. 14.

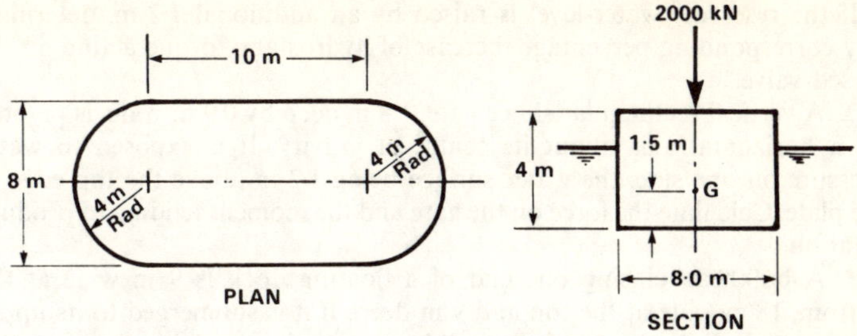

FIG. 14. *See* Example 6

SOLUTION

$$\text{Equivalent mass of superimposed load} = \frac{2 \times 10^6}{9{\cdot}81} = 203\,000 \text{ kg}$$

$$\text{Mass of pontoon} = 90\,000 \text{ kg}$$

$$\text{total} = 293\,900 \text{ kg}$$

$$\text{Volume of water displaced } V = 293{\cdot}9 \text{ m}^3$$

At the water line, $\text{c.s.a.} = 10 \times 8 + 16\pi = 130{\cdot}3 \text{ m}^2$

$$\therefore \quad \text{submerged depth} = \frac{293{\cdot}9}{130{\cdot}3} = 2{\cdot}256 \text{ m}$$

and $\text{height of } B = \frac{1}{2} \times 2{\cdot}256 = 1{\cdot}128 \text{ m}$

$$\therefore \quad BG = 1{\cdot}5 - 1{\cdot}128 = 0{\cdot}372 \text{ m}$$

From Table I, $$I = \frac{10 \times 8^3}{12} + \frac{\pi \times 8^4}{64} = 627{\cdot}7 \text{ m}^4$$

$$\therefore \quad MB = \frac{I}{V} = \frac{627{\cdot}7}{293{\cdot}9} = 2{\cdot}136 \text{ m}$$

deduct $BG = 0{\cdot}372$

and $\text{metacentric height} = 1{\cdot}764 \text{ m}$

QUESTIONS

1. Define the term "centre of pressure" and show how the centre of pressure can be determined for a plane submerged surface.

A reservoir can, if necessary, be emptied through a 1·5-m-diameter horizontal pipeline which discharges freely to the atmosphere. However, this outlet is normally closed by a disc-type valve pivoting about the horizontal diameter.

Determine the torque exerted on the closed valve by the water, when the reservoir water-level is 6·7 m above the valve trunnion-axis.

If the reservoir water-level is raised by an additional 1·2 m, determine any corresponding percentage increase of hydrostatic torque acting on the closed valve.

2. A vertical rectangular sluice gate, 1·8 m deep by 0·9 m wide, is pivoted on a horizontal axis about its centre of gravity. It is exposed to water pressure on one side, the water surface being 1·2 m above the top edge of the plate. Calculate the force on the gate and the moment tending to produce rotation.

3. A bulkhead closing one end of a floating dock is 9 m wide at the bottom, 18 m wide at the top and 9 m deep. If it is submerged to its upper edge, calculate the load on the bulkhead and the depth to the centre of pressure.

4. Water is 18 m deep at the face of a dam, which is vertical to 9 m below water-level and beyond that point is inclined at 30° to the vertical. Specify completely the resultant force on the face of the dam/metre run.

5. A canal of rectangular cross-section is closed by a pair of lock gates, the angle between the gates being 130°.

Each gate is 2·5 m wide and 7·6 m high, and is carried on two hinges, placed 0·3 m from the bottom and 0·3 m from the top respectively. If the upstream depth is 6·4 m and downstream 2·7 m, calculate the force on each hinge.

6. The barge shown in Fig. 15 is to be used to transport a bridge section weighing 5 *MN* as shown. If the barge weighs 50 *MN*, and floats in fresh water, determine:

(*a*) the metacentric height of the unladen barge; and
(*b*) plot a graph showing the variation ot angle of heel with the bridge eccentricity, *e*.

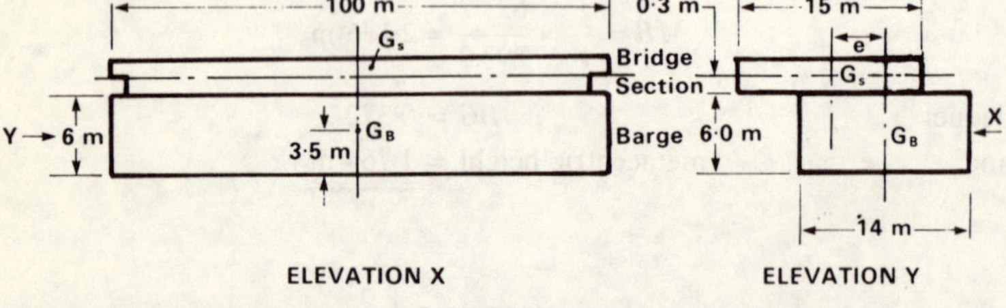

FIG. 15. *See* Question 6

7. A freely floating cylindrical barrier is to be used to hold back fresh water from a spillway outlet crest as shown in Fig. 16.

(*a*) Determine, for no friction between barrier and wall, the required weight per unit length of barrier and the reaction at the wall contact if the barrier is required to float just touching at the crest as shown:
(*i*) with the water level at *C*;
(*ii*) with the water level at *D*.

(*b*) For the weight as determined in (*a*) (*i*) sketch the position of the barrier when the water level falls to that of the crest.

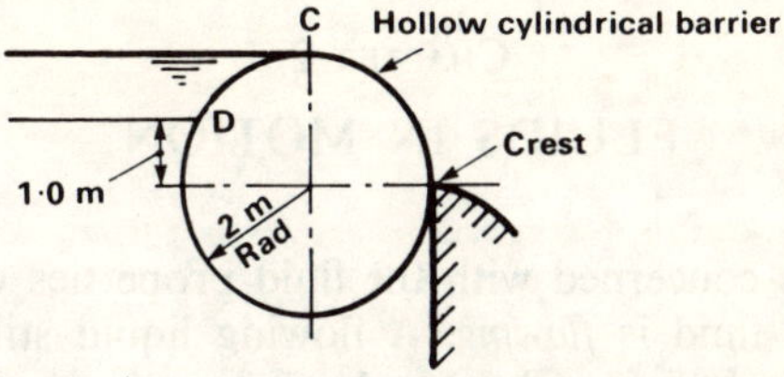

FIG. 16. *See* Question 7

CHAPTER 2

FLUIDS IN MOTION

THIS chapter is concerned with the fluid properties which develop only when the fluid is *flowing*. A flowing liquid still exhibits the properties described in Chapter 1 of course, but in some circumstances the values of these properties may require calculation by a method differing from that given in Chapter 1. For example, the density of a *liquid* is not normally affected by motion, the density of a *gas* however may be affected. Also pressure in moving fluids may differ from hydrostatic pressure. The remaining chapters of this book deal mainly with fluids of motion; hence it is essential to understand the fluid properties which exist when the fluid is in motion and also the different types of fluid flow.

In a *moving* fluid, each element of fluid experiences stresses exerted on it by other elements of the fluid which surround it. The stress at each part of the surface of an element is resolved into components normal and tangential to the direction of the fluid motion known respectively as *pressure* (defined in Chapter 1) and shear stress. Pressures are exerted whether the fluid is moving or at rest, but shear stresses occur only in moving fluids. The property which gives rise to shear stresses is called *viscosity*.

VISCOSITY

Fig. 17 (*a*) depicts a large, flat, floating board being pulled slowly at speed V by a force F over the surface of a liquid contained in a tank. The movement of the board causes the liquid to flow in the

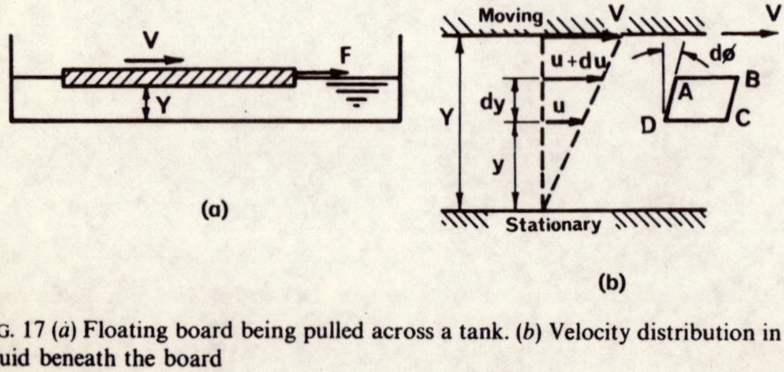

FIG. 17 (*a*) Floating board being pulled across a tank. (*b*) Velocity distribution in the liquid beneath the board

same direction as the board with a velocity which varies linearly from 0 at the bottom of the tank to V, where it is in contact with the board, as shown in Fig. 17 (*b*). Assuming the velocity distribution is the same throughout all the liquid beneath the board, the force required to move the board at the steady speed V is given by

$$F = \mu A . \mathrm{d}u/\mathrm{d}y$$

and the force-resisting motion, which is equal and opposite to F (if V = constant), is the *viscous drag*,

$$R = \mu A . \mathrm{d}u/\mathrm{d}y$$

In these equations:

> μ (mu) is the *coefficient of dynamic viscosity* or *coefficient of molecular viscosity*
> A is the area of one side of the board;
> $\mathrm{d}u/\mathrm{d}y$ is the velocity gradient across the liquid; and
> fluids which behave in this way are called *Newtonian fluids* in contrast to *non-Newtonian* fluids which do not.

It should also be noted that, for flow between plane boundaries, the velocity gradient is given by $\mathrm{d}u/\mathrm{d}y = V/Y$, where

> V = the speed of one boundary relative to the other, and
> Y = distance between the boundaries.

No proof of the equation $R = \mu A . \mathrm{d}u/\mathrm{d}y$ is offered, as it may be thought of as a definition of μ. It could be verified experimentally by measuring the forces necessary to pull boards of various sizes at different speeds across a given liquid. The depth Y could also be varied. It would be found that $F \propto A$ when V and Y are kept constant, and $F \propto V/Y$ when A is kept constant. From these it can be concluded that $F \propto AV/Y$ and μ can be defined as the constant of proportionality. In such an experiment the dimensions of each board must be large compared with the distance Y, for otherwise the accuracy of the results would be impaired by the undesirable edge effects. By this is meant the velocity distribution near the edges, which is different from, and more complex than, the velocity distribution in the liquid under the central part of the board; the area (near the edges) where this occurs will be a smaller proportion of the whole area of a large board than it would of smaller boards.

Notes on viscosity

1. Viscous flow. In many engineering problems the effects of viscosity on the flow are small, and are therefore ignored. The liquid is then described as *inviscid* or, sometimes, *ideal*, and μ is taken as zero. If, however, the term *viscous flow* is used, this means that viscosity is not to be ignored.

2. *Velocity.* In viscous flow a fundamental rule is that the fluid velocity at the boundary must be the same as the velocity of the boundary, as it is in Fig. 17 (*b*). Otherwise, there is an infinite velocity gradient next to the boundary and, since $R = \mu A . du/dy$, an infinite shear stress (*see* note 3).

3. *Shear strength.* It is well known that liquids *at rest* possess no shear strength, for in bulk they deform to fill their container, whatever its shape. When in motion, however, they do possess shear strength, because if R is the viscous drag developed over an area A the shear stress is $R/A = \mu du/dy$. Referring to Fig. 17 (*b*), if an element deforms from a rectangular shape to the parallelogram $ABCD$ in time dt, then $d\phi/dt = du/dy$, and $R/A \propto d\phi/dt$. In liquids, therefore, shear stress is proportional to the *rate* of shear strain. It will be remembered that, for elastic solids, shear stress $\propto$ shear strain.

4. *Dimensions of μ.* The conception of *dimensions* is introduced on p. 67, Chapter 5, and the reader who is not already familiar with the idea of dimensions would be well advised to refer to p. 67 before proceeding, as this note concerns the dimensions of μ. Since viscous drag, $R = \mu A . du/dy$, μ has the dimensions of a stress (R/A) divided by a velocity gradient; *i.e.*

$$(MT^{-2}L^{-1}) \div (LT^{-1}/L) = ML^{-1}T^{-1}$$

Hence, in the m kg s system, the units of μ are kg/(ms).

An alternative way of describing these units is obtained by expressing μ in the form $\mu = R/A . du/dy$), from which they can be defined as N s/m^2, *i.e.*:

$$1 \text{ kg/ms} \equiv 1 \text{ Ns/m}^2$$

In the c.g.s. system the unit of μ is the *poise*, which is the same as 1 g/(cm sec). Hence 1 kg/ms = 10 poise = 1000 centipoise.

5. *The coefficient of kinematic viscosity*, ν (*nu*), is defined by

$$\nu = \mu/\rho$$

ν is measured either in m^2/s or in Stokes, 1 Stoke being 1 cm^2/s, and the relationship between these units is:

$$1 \text{ centistoke (cSt)} = 10^{-6} \text{ m}^2/\text{s}$$

where 1 Stoke = 100 centistokes.

6. *Viscous drag.* The formula $R = \mu A . du/dy$ was associated on p. 23 with the motion of one *plane* relative to another. The same formula can be applied to the relative motion of two concentric cylinders of nearly equal diameters with liquid between. This is quite a common engineering arrangement, occurring, for example, where a shaft, lubricated by oil, turns in its bearing. Fig. 18 shows how the concentric cylinders can be approximated to a series of parallel

planes when the distance between them is small compared with the mean radius.

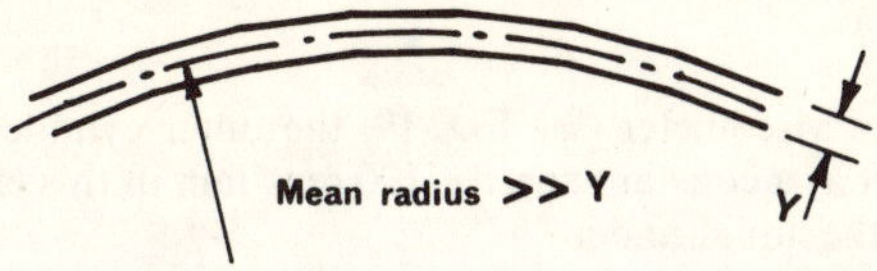

FIG. 18. Flat-sided figures approximating to cylinders

Measurement of viscosity

A device for the measurement of viscosity is called a *viscometer*. A number of different types of viscometer exist, but, as the purpose of this book is to establish certain principles of hydraulics rather than to describe hydraulic machinery and apparatus, these should be read up elsewhere. It is convenient, however, to mention three ways in which μ can be determined. They are:

(*a*) *By means of a torsion viscometer—see* Example 7, which illustrates how the formula $R=\mu A.\mathrm{d}u/\mathrm{d}y$ is applied to concentric cylinders.

(*b*) *By measurement of the pressure drop in pipe-flow.* This method is again referred to on p. 33.

(*c*) *By Stokes' law for a falling sphere.* Stokes' formula is:

$$F=3\pi\mu \mathrm{d}u$$

where F is the drag experienced by a very small sphere of radius r falling freely through a liquid of viscosity μ with a speed u. Stokes' formula is valid only if the Reynolds number for the flow is less than (about) 1, the Reynolds number being defined by

$$R_e=ud/v$$

where d is the diameter of the sphere. In other words, Stokes' formula is valid only at very small velocities—but just how small depends also on v and d.

The significance, illustrated by Example 8, of the critical Reynolds number, $R_e = 1$, is that at $R_e < 1$ the flow past the sphere is *viscous* and the resistance to motion is *viscous drag*, whereas at $R_e \gg 1$ the flow past the sphere is *turbulent* and the resistance to motion is a mixture of *friction* and *form drag* due to turbulent flow. The difference between viscous flow and turbulent flow, in connection with pipe-flow, is discussed at greater length on p. 27.

EXAMPLE 7

If the liquid is 130 mm deep in a torsion viscometer and the cylinders are 150·5 mm and 150 mm diameter respectively, what is the viscosity of a liquid which produces a torque of 540 mN m upon the inner cylinder when the outer one rotates at 100 revs/min?

SOLUTION

In this type of viscometer (*see* Fig. 19) the outer cylinder is rotated by turning the shaft at a constant speed—100 revs/min in this example. Owing to its viscosity, the liquid under test transmits a torque to the inner cylinder, which is suspended by a torsion wire fixed at its upper end. This wire forms part of the torsion head; the angle through which the wire is twisted is proportional to the torque applied to it. Consequently, if the torsion head dial is calibrated in torque units the pointer attached to the torsion wire gives a direct reading of the transmitted torque.

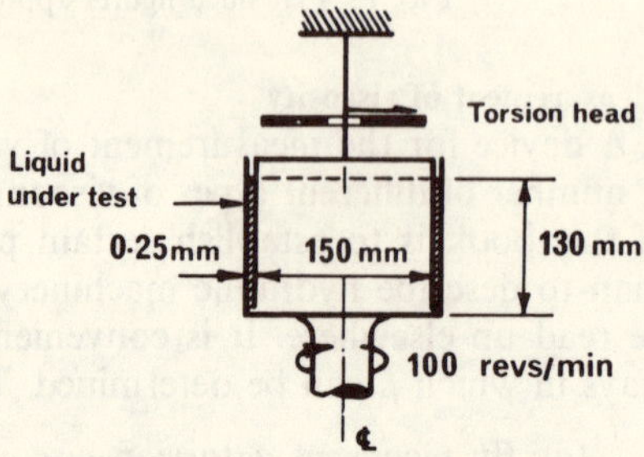

FIG. 19. One kind of torsion viscometer

In this example the torque = 0·54 N m

i.e. $0{\cdot}54 = \text{Viscous drag } (R) \times 0{\cdot}075$

$\therefore\ R = 7{\cdot}2 \text{ N}$

This is the drag on the surface of the inner cylinder, whose area is

$$A = 0{\cdot}130 \times \pi \times 0{\cdot}150 = 0{\cdot}0613 \text{ m}^2$$

$$\therefore \text{ Viscous stress} = R/A = 7{\cdot}2/0{\cdot}0613 = 117{\cdot}5 \text{ N/m}^2$$

Since the outer cylinder rotates at 100 revs/min, its tangential velocity is $100/60 \times 0{\cdot}1505\pi = 0{\cdot}788$ m/s.

Hence, velocity gradient, $du/dy = 0{\cdot}788/0{\cdot}00025 = 3152 \text{ s}^{-1}$

From the formula, $R = \mu A . du/dy$

$$\begin{aligned}\mu &= \text{viscous stress} \div \text{velocity gradient}\\ &= 117{\cdot}5/3152 = 0{\cdot}0373 \text{ N s/m}^2\\ &= \underline{\underline{37{\cdot}3 \text{ mN s/m}^2}}\end{aligned}$$

(It should be noticed that, as the velocity gradient has units of s^{-1} and the viscous stress has units of N/m^2, the viscosity μ has units of N s/m^2.)

VISCOSITY IN PIPE-FLOW

In a famous experiment, first performed by Osborne Reynolds (1842–1912) and since performed by others with more refined apparatus, great care was taken to obtain a flow of water under steady conditions in a glass tube. A number of tests were performed at various rates of discharge from the same tube; in each a thin stream of dye was introduced into the flow and its behaviour observed.

Fig. 20 (*a*) shows the stream of dye at low velocities and Fig. 20 (*c*) shows what happens to it at the highest velocities. Fig. 20 (*b*) indicates the transition phase between the other two, a transition

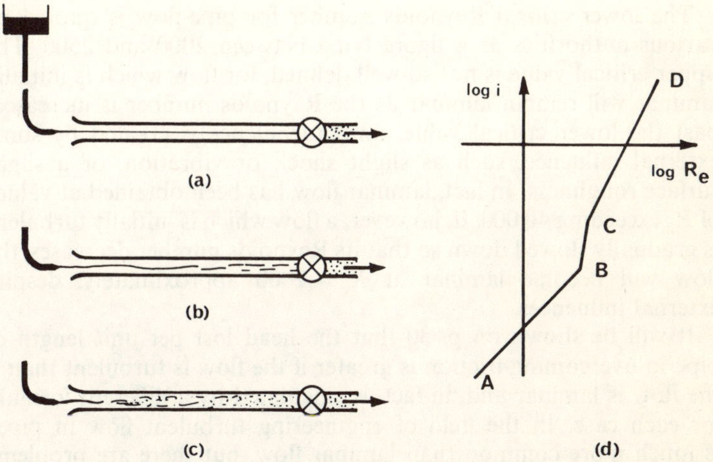

FIG. 20. The results of Reynolds' famous experiments are indicated: (*a*) for laminar flow, (*b*) during the transition stage, and (*c*) for turbulent flow. (*d*) The graph of log R_e against log *i*; *AB* is laminar flow, *CD* is turbulent flow and *BC* is the transition stage

which occurs over a very limited range of velocity. Over the years the experiment has been repeated with a variety of liquids instead of water flowing through tubes of a great many different diameters.

The results of all these tests are truly significant only when they are expressed in terms of the *Reynolds number*, R_e, which is defined for pipe-flow by

$$R_e = \rho ud/\mu = ud/v$$

Since the dimension of v are L^2/T, the dimensions of u are L/T and the dimension of d is L, it can easily be verified that R_e is dimensionless. *Hydraulic gradient*, *i*, is another dimensionless quantity and so is (the more familiar) *percentage*. Just as the statement "all authors should receive 10 per cent increase in royalties" means something specific regardless of whether the author is successful or not, and regardless of whether his books are long or short, so a suitable dimensionless number can often be used to describe a flow phenomenon which is true regardless of scale (*i.e.* the size of the apparatus) and regardless of the fluid's properties, such as density and viscosity. Two such observations can be made about flow in circular pipes:

1. If R_e for a pipe-flow is less than the *lower critical Reynolds*

number the flow will be *laminar*, alternatively described as *viscous*.

2. If R_e for the pipe-flow exceeds the *upper critical Reynolds number*, the flow will be *turbulent*.

The lower critical Reynolds number for pipe-flow is quoted by various authorities as a figure lying between 2000 and 2500. The upper critical value is not so well defined, for flow which is initially laminar will remain laminar as the Reynolds number is increased, past the lower critical value, until turbulence is created by some external influence, such as slight shock or vibration, or a slight surface roughness. In fact, laminar flow has been obtained at values of R_e exceeding 40,000. If, however, a flow which is initially turbulent is gradually slowed down so that its Reynolds number decreases, the flow will become laminar at $R_e = 2000$ approximately, despite external influences.

It will be shown on p. 30 that the head lost per unit length of pipe in overcoming friction is greater if the flow is turbulent than if the flow is laminar and, in fact, is calculated by a different formula for each case. In the field of engineering turbulent flow in pipes is much more common than laminar flow, but there are problems in which the calculation of Reynolds number is a necessary preliminary to the main calculation, *e.g.* to decide by which formula to calculate the friction head lost.

Again, flow past a sphere can be turbulent or laminar. As shown in Example 8, the drag on the sphere can be calculated by Stokes' formula only if it is viscous (*i.e.* laminar).

In connection with Example 8, moreover, it should be realised that the movement of a sphere through stationary fluid is, hydrodynamically, identical to flow past a stationary sphere, since the velocity of the sphere relative to the fluid is the significant velocity.

EXAMPLE 8

Distinguish clearly between the following types of fluid-flow: non-uniform; turbulent; and unsteady.

A hailstone, with diameter 5 mm and density 920 kg/m^3, falls through the atmosphere and into an open tank of oil, with density 810 kg/m^3. Determine the rate of descent in air, density 1·3 kg/m^3, assuming a coefficient of drag of 0·45.

Compare the rate of descent in air with that in the oil if the latter has a coefficient of dynamic viscosity of 0·27 N s/m^2.

SOLUTION

The flow in a *region* is said to be uniform if the fluid velocity does not vary from *place to place*. NON-UNIFORM describes flow in a region where this is not the case.

Flow *at a point* (in space) is said to be steady if the velocity there does not vary with *time*. Otherwise it is UNSTEADY.

(Alternatively, for readers who are familiar with partial differentials and who understand the notion of partial differentiation, these terms can be defined mathematically thus:

A point in space (which is 3-dimensional) can be described by the co-ordinates x, y, z, just as a point on a graph (which is 2-dimensional) can be described by the co-ordinates x, y. Since the velocity of flow may vary as x, y and z vary, and since it may also vary with time, we say velocity is a function of x, y, z and t, *i.e.*:

$$u = f(x,y,z,t)$$

then:

for *steady* flow, $\partial u/\partial t = 0$;
for *unsteady* flow, $\partial u/\partial t \neq 0$;
for *uniform* flow, $\partial u/\partial x$, $\partial u/\partial y$, $\partial u/\partial z = 0$; but
for *non-uniform* flow, at least one of $\partial u/\partial x$, $\partial u/\partial y$, $\partial u/\partial z$ is not zero.

Strictly, TURBULENT flow is both *non-uniform* and *unsteady*, although the reverse is not necessarily true. In it the flow velocity at any point is continuously changing with small haphazard fluctuations superimposed on the main flow velocity. *See* also Chapters 10 and 11.

An example of *steady uniform* flow is provided by water flowing through a pipe of constant cross-section, and running full, the water having a constant velocity—for example, 0·15 m/s. If the water in the pipe were to be accelerated by means of, say, a piston near one end the flow would still be *uniform*, but would be *unsteady* by virtue of the acceleration. When slow flow, at a constant rate of discharge, occurs in a diverging or converging pipe it will be *steady* but *non-uniform*, because the velocity at any point in the pipe does not change with time, whereas, owing to the change of cross-sectional area, the velocity does vary from section to section.

As an introduction to the calculation, the reader is reminded that a sphere falling under its own weight through a fluid will attain a constant velocity, called the *terminal velocity*, when the drag force is equal and opposite to its weight, so that the net force acting on the sphere is then zero and, by Newton's second law, the acceleration is also zero. If $R_e < 1$ the drag force is entirely viscous and the drag is determined by Stokes' law. If $R_e \gg 1$ the resistance is a complicated mixture, *see* p. 173, and may be calculated from:

$$C_D = D/(\tfrac{1}{2}\rho U^2 A)$$

where C_D = coefficient of drag;
D = drag force;
ρ = density of the fluid;
U = free stream velocity;
= velocity of sphere relative to the fluid;
A = maximum cross-sectional area of the object across the flow path.

Form drag, it should be noted, is a resistance which depends on the shape of the body and which is associated with the loss of energy in the eddies which form behind the body.

For the hailstone in air, drag = Weight of hailstone

$$= \frac{4}{3} \times \pi \times \left(\frac{5}{2}\right)^3 \times 10^{-9} \times 920 \times 9{\cdot}81$$
$$= 0{\cdot}59 \times 10^{-3}\ \text{N}$$

Also, $$\text{Drag} = C_D \times \tfrac{1}{2}\rho U^2 A$$

i.e. $$0{\cdot}59 \times 10^{-3} = 0{\cdot}45 \times \tfrac{1}{2} \times 1{\cdot}3 \times \frac{\pi}{4} \times 5^2 \times 10^{-6} \times U^2$$

from which $U^2 = 102{\cdot}8$, and

$$U = \underline{\underline{10{\cdot}14\ \text{m/s}}}$$

In oil, assume Stokes' formula applies, so that

$$\text{Drag. } D = 3\pi d\mu u$$

Also, D = Weight of hailstone, allowing for buoyancy

$$= 4/3\pi \times 2{\cdot}5^3 \times 10^{-9} \times (920 - 810) \times 9{\cdot}81$$
$$= 70{\cdot}6 \times 10^{-6}\ \text{N}$$
$$\therefore\ 70{\cdot}6 \times 10^{-6} = 3\pi \times 5 \times 0{\cdot}27 \times 10^{-3} u$$

Hence,

$$u = \underline{\underline{5{\cdot}55 \times 10^{-3}\ \text{m/s}}}$$

i.e. $$u = 5{\cdot}55\ \text{mm/s}$$

Having calculated *u*, we can now check that Stokes' formula is valid by evaluating the Reynolds number.

$$R_e = \rho u d/\mu = 810 \times 5{\cdot}55 \times 10^{-3} \times 5 \times 10^{-3}/0{\cdot}27$$
$$= 0{\cdot}083 \ll 1$$

Darcy–Weisbach formula for pipe-flow

h_f, the head lost in overcoming friction in a pipe of diameter d and length l, depends on the velocity of flow, u, and on a coefficient of friction, f. It is given by the empirical Darcy–Weisbach formula (often referred to simply as the Darcy formula), namely,

$$h_f = 4flu^2/(2gd)$$

Although it is tempting to simplify the equation by dividing the 2 into the 4, this is not usually done because the 2 belongs to one familiar group, $u^2/(2g)$, and the 4 to another, the *hydraulic mean depth* $d/4$ (*see* p. 167).

Fig. 21 indicates how f varies with the Reynolds number, R_e, for rough and smooth pipes a subject dealt with at greater length in Chapter 11. The relative roughness of a pipe can be expressed as k/r, k being the mean height of projection and r the radius of the pipe. In turbulent flow, as might be expected, the greater values of f correspond to the greater values of k/r. In viscous flow, however, when $R_e < 2000$, f is independent of the pipe roughness. In this range it is found that $f = 16/R_e$ fits the straight-line graph. If $16/R_e$ is substituted for f in the Darcy formula, in fact, it will be found to yield Poiseuille's formula, which is derived in Example 9.

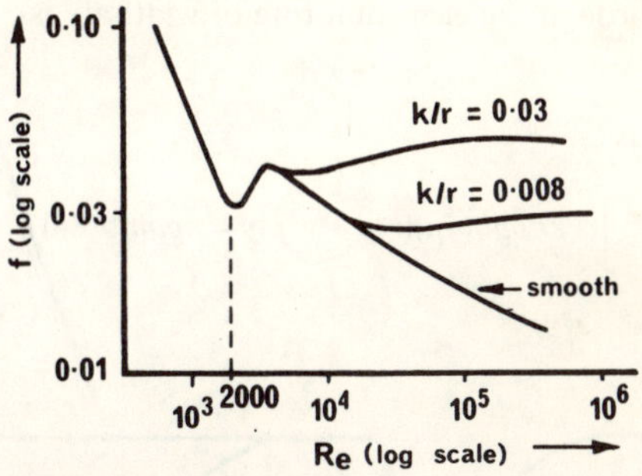

FIG. 21. Graph of friction coefficient, f, against Reynolds number (log scales). Based on experiments by Nikuradse

EXAMPLE 9

Define the coefficient of (dynamic) viscosity, μ. Show that the mean velocity, $\bar{u}$, of laminar flow in a circular pipe of uniform diameter d is

$$\bar{u} = \tfrac{1}{32} \cdot \rho g i d^2/\mu$$

where ρ is the density of the fluid and i the energy gradient.

SOLUTION

$$\text{Energy gradient, } i = h_f/l$$

where h_f is the friction head lost in a length l of pipe. By putting $l = 1$ in this equation, it is seen that i can also be interpreted as the loss of head per unit length of pipe, and therefore the pressure drop per unit length $= \rho g i$. Considering the liquid "core" of radius y (Fig. 22 (a)), the driving force causing its movement is:

Force = $$\text{Pressure drop} \times \text{c.s.a.} = \rho g i \times \pi y^2$$

For steady flow, the driving force is equal and opposite to the force resisting motion, namely the viscous shear force on the boundary of the core, $\mu A . \mathrm{d}u/\mathrm{d}y$. In this expression A is the surface area of the core, *i.e.*:

$$A = 2\pi y \times \text{Unit length}$$

Hence, inserting a minus sign because the forces act in opposite directions,

$$-\mu . 2\pi y . \mathrm{d}u/\mathrm{d}y = \pi y^2 \rho g i$$

i.e. $$-u = \rho g i/(2\mu) \int y . \mathrm{d}y = \rho g i/(4\mu) . y^2 + C$$

Since $u = 0$ when $y = r$, the constant of integration $C = -\rho g i r^2/(4\mu)$.

Hence

$$u = \rho g i/(4\mu).(r^2 - y^2) \qquad (1)$$

The maximum velocity occurs when $y = 0$ and is

$$u_{\max} = \rho g i r^2/(4\mu)$$

The discharge through the elemental ring of width dy is

$$u \times 2\pi y.\mathrm{d}y$$

∴ total discharge is

$$Q = \int_0^r 2\pi\rho g i/(4\mu)(r^2 y - y^3)\mathrm{d}y = \pi\rho g i r^4/(8\mu)$$

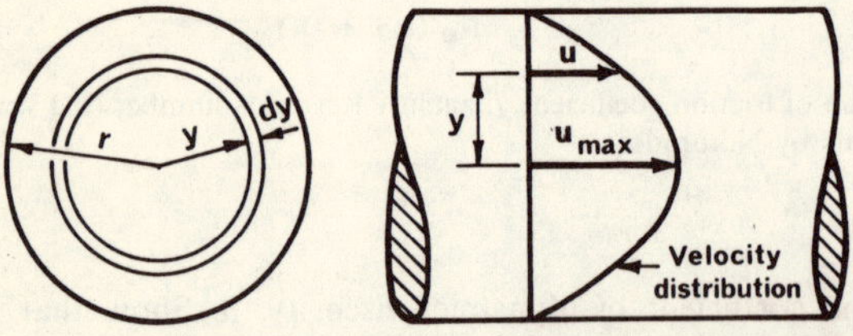

(*a*) (*b*)

FIG. 22. *See* Example 9

The main velocity, $\bar{u} = Q/(\pi r^2) = \rho g i r^2/(8\mu) = \frac{1}{2}u_{\max}$

∴ since $r = d/2$,

$$\underline{\bar{u} = \frac{1}{32}.\rho g i d^2/\mu}$$

which is one form of Poiseuille's equation.

Rewriting this equation in its more usual form,

$$i = 32\mu\bar{u}/(\rho g d^2)$$

and head lost in friction, $h_f = il$,

hence: $h_f = 32\mu\bar{u}l/(\rho g d^2)$

(It should be noticed that, from equation (1), the velocity distribution shown in Fig. 22 (*b*) is paraboloidal.)

Poiseuille's formula for pipe-flow

It has been shown in Example 9 that Poiseuille's formula for laminar flow in pipes is $i = 32\mu\bar{u}/(\rho g d^2)$, which can also be written in the form

$$h_f = 32\mu\bar{u}l/(\rho g d^2)$$

where h_f is the head lost in overcoming friction in a length l of pipe. For any particular liquid, ρg is known. h_f can be measured by connecting pressure tappings, a distance l apart, to the limbs of a sensitive pressure gauge. By measuring the discharge from the pipe under test and dividing by $\pi d^2/4$, the mean velocity, $\bar{u}$, can be calculated, if the diameter, d, of the pipe is known. Hence, slow flow through a pipe provides a suitable method of determining μ, the apparatus being as shown in Fig. 23.

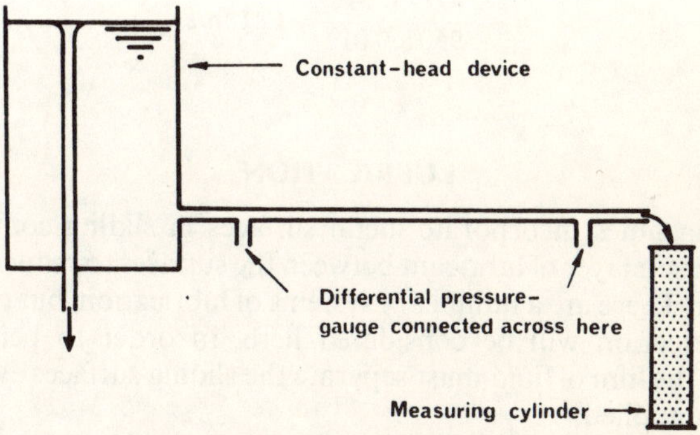

FIG. 23. Apparatus for determining μ

EXAMPLE 10

In a laboratory test, oil of density 950 kg/m³ is pumped through a horizontal perspex tube of 10 mm diameter. Piezometer readings show that a head loss of 84 mm occurs in the test length of 2·5 m when the flow rate through the tube is $1{\cdot}54 \times 10^{-3}$ litres/s. Determine:

(*a*) the oil viscosity in (*i*) centipoise, (*ii*) centistokes;
(*b*) the Reynolds number for this flow rate;
(*c*) the lower critical velocity of this oil in this particular tube.

SOLUTION

Assume the flow is laminar so that Poiseuille's equation applies. Since $1{\cdot}54 \times 10^{-3}$ litres/s $= 1{\cdot}54 \times 10^{-6}$ m³/s,

$$\bar{u} = \frac{Q}{A} = 1{\cdot}54 \times 10^{-6} \div \frac{\pi}{4} \times 0{\cdot}01^2 = 0{\cdot}0196 \text{ m/s}$$

$$\mu = \frac{\rho g d^2 h_f}{32 l \bar{u}} = \frac{950 \times 9{\cdot}81 \times 0{\cdot}01^2 \times 0{\cdot}084}{32 \times 2{\cdot}5 \times 0{\cdot}0196}$$

$$= 0{\cdot}0499 \text{ kg/ms}$$

Then,

$$R_e = \frac{\rho \bar{u} d}{\mu} = \frac{950 \times 0{\cdot}0196 \times 0{\cdot}01}{0{\cdot}0499} = 3{\cdot}73$$

Hence,

$$\mu = 0{\cdot}0499 \times 1000 = \underline{49{\cdot}9 \text{ centipoise}}$$

and

$$\nu = \frac{\mu}{\rho} = \frac{0{\cdot}0499}{950} = 5{\cdot}25 \times 10^{-5} \text{ m}^2/\text{s}$$

$$= \underline{52{\cdot}5 \text{ centistokes}}$$

(*c*) Taking the lower critical value of Reynolds number as 2300, a value often used,

$$\bar{u} = \frac{2300 \times 0{\cdot}0499}{950 \times 0{\cdot}01} = 12.1 \text{ m/s}$$

LUBRICATION

When machines incorporate metal surfaces in sliding contact it is usual to use a layer of lubricant between the surfaces to reduce friction and wear. There are a number of systems of lubrication, but only fluid film lubrication will be considered here. In order to perform its function the film of fluid must separate the sliding surfaces even when a load is applied.

Consider first a pair of parallel surfaces separated by a layer of fluid. If a perpendicular load is applied the fluid will be "squeezed" out, since there is no hydraulic pressure to enable the fluid to resist the squeezing action. One way of overcoming this problem for sliding surfaces is to set the surfaces at some angle θ (Fig. 24). The difference of area between inlet and outlet sets up a velocity distribution which

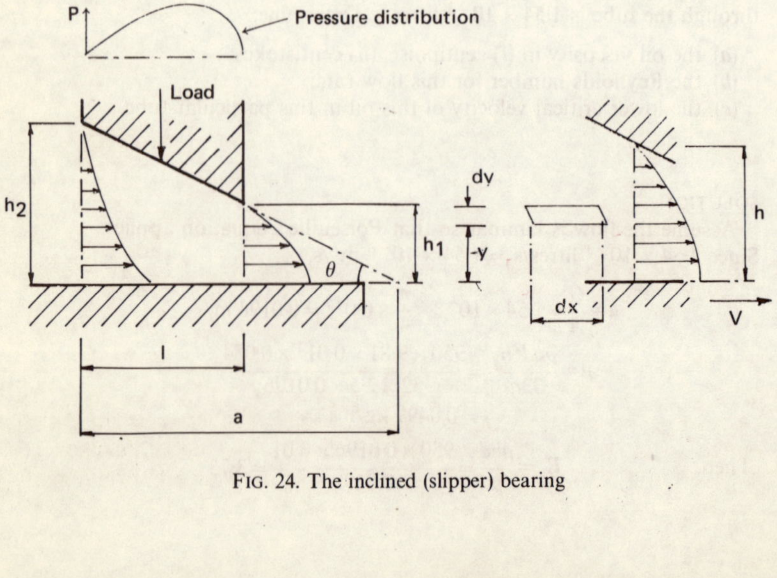

FIG. 24. The inclined (slipper) bearing

varies with x and this leads to a build up of pressure which enables the lubricant to resist a perpendicular load. This mechanism is now analysed.

The bearing illustrated is assumed to be of unit width, and moving with uniform velocity V. If flow is viscous, then pressure force must balance viscous force on any element of fluid:

$$\mathrm{d}p \cdot \mathrm{d}y = -\mathrm{d}\tau . \mathrm{d}x \quad \therefore \frac{\mathrm{d}p}{\mathrm{d}x} = \frac{-\mathrm{d}\tau}{\mathrm{d}y}$$

and since $\tau = -\mu \dfrac{\mathrm{d}u}{\mathrm{d}y} \quad \therefore \dfrac{\mathrm{d}p}{\mathrm{d}x} = \mu \dfrac{\mathrm{d}^2 u}{\mathrm{d}y^2}$

Re-arranging and integrating:

$$u = \frac{1}{\mu}\left(\frac{\mathrm{d}p}{\mathrm{d}x}\right)\frac{y^2}{2} + C_1 y + C_2$$

Now $u \to V$ as $y \to 0$, and $u \to 0$ as $y \to h$.

Hence constants C_1 and C_2 may be evaluated and the equation becomes:

$$u = \frac{1}{2\mu}\left(\frac{\mathrm{d}p}{\mathrm{d}x}\right)(y^2 - hy) + V - \frac{Vy}{h}$$

The discharge Q through the bearing is given by:

$$Q = \int_0^h u\mathrm{d}y = \frac{Vh}{2} - \frac{h^3}{12\mu}\left(\frac{\mathrm{d}p}{\mathrm{d}x}\right)$$

Hence $\mathrm{d}p = 12\mu\left(\dfrac{V}{2h^2} - \dfrac{Q}{h^3}\right)\mathrm{d}x$

substituting $h = (a-x)\tan\theta$ and integrating:

$$p = \frac{6\mu x}{a(a-x)\tan^2\theta}\left(V - \frac{Q(2a-x)}{a(a-x)\tan\theta}\right)$$

but $p \to 0$ as $x \to 0$ and as $x \to l$

$$\therefore p = \frac{6\mu V x(l-x)}{(2a-l)(a-x)^2\tan^2\theta}$$

The total perpendicular load which can be sustained by the bearing can be obtained by integration.

$$\int_{a-l}^{a} p\mathrm{d}x = \frac{6\mu V}{\tan^2\theta}\left(l_n\left(\frac{a}{a-l}\right) - \frac{2l}{2a-l}\right)$$

A journal bearing is a bearing (internal diameter $=D$) which surrounds a revolving shaft (diameter $=d$). Lubricant is used to separate the surfaces and therefore fills the clearance between shaft and bearing. To apply the above theory it must be assumed that the shaft runs eccentrically in the bearing to form a "wedge" of lubricant. The theory assumes that flow is laminar and that viscosity is constant. In practice it is possible for turbulent flows to be encountered at high speeds, and the viscosity will certainly *not* remain constant due to the heat generated through friction.

The above approach should be used if interest is centred on the load-carrying capacity of a lubricated bearing. On the other hand, if an estimate of the viscous resistance to motion is all that is required, it is often possible to obtain satisfactory estimates based on the assumption of uniform film thickness (*see* Example 7).

POWER

Energy, power and "work done" are familiar quantities to the student of applied mechanics, but perhaps it would be as well to revise the relationships between them. These are:

(*a*) *Work done = force × distance moved in the direction of the force.* To be more precise, work is done on an object by a force which moves the object. If it is free to do so, the object will move in the direction of the force, and the work done on it will be the product of the force and the distance moved by the object; if it is constrained to move in some other direction the work done on it is the product of the force and the component, in the direction of the force, of the distance moved by the object. Moreover, by doing work on the object the driving agent loses energy and, in fact, *work done* is merely another name for this loss of energy. Both are measured in the same units, commonly Newton metres, N m. Hence:

(*b*) *Work done = loss of energy by the driving agent.*

Rate of doing work is termed *power* and is measured in N m/s or watts. Hence:

(*c*) *Power = Rate of doing work*
= Work done per second
= Force × distance moved per second
= Force × velocity
(again, in the direction of the force)
= Energy lost (by the driving agent) per second.

Flow from the pipe outlet in Fig. 25 lowers the surface of the liquid in the tank. When the outlet is a depth H below the surface, therefore, and 1 kg of liquid escapes past the valve the potential energy of the liquid in the tank is reduced by gH Newton metres. Further, if the rate of discharge is Q m^3/s, ρQ kg of liquid flow past the valve each second and when the surface is H metres above the outlet the potential energy of the liquid in the tank is reduced at the rate of ρgQH Newton metres per second. Assuming there is no loss of energy in the short outlet shown, the flowing liquid possesses kinetic energy equal to the potential energy lost by the liquid in the tank, and is therefore capable of doing work at the rate of ρgQH watts.

For example, for water flowing at the rate of $3{\cdot}5 \times 10^{-4}$ m^3/s under a head of 3 m,

$$\begin{aligned} \text{power available} &= 1000 \times 9{\cdot}81 \times 3{\cdot}5 \times 10^{-4} \times 3 \\ &= 10{\cdot}3 \text{ N m/s} \\ &= 10{\cdot}3 \text{ watts} \end{aligned}$$

In a long pipe-outlet some of the kinetic energy is used in overcoming friction, but the same expression, ρgQH, for power is still true at every section of the outlet pipe, providing H is now interpreted as the total head of the section.

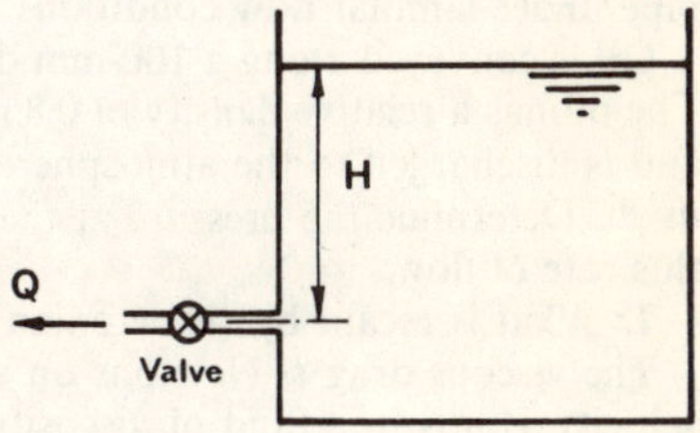

FIG. 25. Water flowing from a tank, giving a power of ρgQH

QUESTIONS

1. What is meant by the *viscosity* of a fluid and how is the *coefficient of dynamic viscosity* defined?

Obtain an expression for the loss of head which results from the flow of a viscous fluid through a circular-section pipeline, in terms of the length of pipeline l, the internal diameter of pipeline d, the mean velocity of flow u, the fluid specific weight w, and the coefficient of dynamic viscosity μ.

How, and in what circumstances, may the Darcy and Poiseuille equations, for resistance to flow through circular-section pipelines, be related?

2. Using the Darcy formula for the head lost in pipe friction,

$$h_f = 4flu^2/(2gd)$$

calculate the rate of flow of gas through a pipe of 200 mm diameter and of length $l = 150$ m between two gas holders. A water manometer at the pipe inlet reads 130 mm, and another manometer at the outlet, placed 30 m higher, 50 mm. The densities of the gas, air and water are 0.72, 1·3 and 1000 kg/m^3 respectively, and the friction coefficient, $f = 0{\cdot}005$.

3. A shaft, 100 mm diameter, runs in a bearing of length 200 mm. The two surfaces are separated by a film of oil. The thickness of the oil film is 2·5 mm and the viscosity of the oil is 153 mN s/m^2. Calculate the torque necessary to rotate the shaft at 30 revs/min.

4. The space between two large flat and parallel walls, 25 mm apart, is filled with a liquid of absolute viscosity 700 mN s/m^2. Within this space a thin flat plate, 250 mm × 250 mm × 2 mm thick, is towed at a velocity of 150 mm/s at a distance of 8 mm from one wall, the plate and its movement being parallel to the walls. Assuming linear variations of velocity between the plate and the walls, determine the force exerted by the liquid on the plate.

5. Explain what is meant by the term "coefficient of viscosity." The end-thrust of a vertical shaft is taken by a flat footstep bearing 150 mm in diameter. The surfaces of the two elements of the bearing are separated by a film of oil 0·25 mm thick. The viscosity of the oil is 150 mN s/m^2, and the shaft rotates at 2000 revs/min. Calculate the power absorbed by fluid friction. Work from first principles.

[*Hint:* Develop an expression for the torque required for an elemental ring of the bearing and integrate from $r=0$ to $r=75$ mm.]

6. Derive an expression for the quantity of fluid flowing through a circular pipe under laminar flow conditions.

Oil is conveyed along a 100-mm diameter pipe which is 600 m in length. The oil has a relative density of 0·85, an absolute viscosity of 150 mN s/m^2 and is discharged to the atmosphere from the pipe outlet at the rate of 30 m^3/h. Determine the pressure at the pipe inlet and the power required for this rate of flow.

7. What is meant by "Newtonian flow"?

The viscous drag D Newtons on a sphere of radius R metres moving at velocity V m/s in a fluid of viscosity μ kg/ms is given by the relationship $D = 6\pi R\mu V$ provided that Newtonian flow conditions apply.

At a given time a lake of maximum depth 3 m is charged with silt of mean particle diameter 0·005 mm and relative density 2·5. Estimate the time it would take for the lake water to become completely clear. Assume a water viscosity of $1{\cdot}8 \times 10^{-3}$ kg/ms. What are the probable reasons for inaccuracy in this estimate?

8. A horizontal pipeline is 200 m long and has a diameter of 0·15 m. It conveys a light machine oil of relative density 0·907 and viscosity 0·114 kg/ms. A pressure gauge at the upstream end of the pipe records a pressure of 100 kN/m^2 gauge. A second pressure gauge is fitted upstream of a flow control valve at the outlet. The reading on this gauge depends on the rate of flow. Determine:

(*a*) the maximum flow rate if the mode of flow is to be laminar;
(*b*) the reading on the second pressure gauge at this rate of flow.

Assume the lower criticial Reynolds number to be 2300.

9. In a hydro-electric scheme the turbines are supplied by three parallel pipes each of diameter 1·5 m and length 2 km. They connect a reservoir, water surface level 535·00 m *AOD*, to a turbine intake chamber where the pipe centre-line level is 472·00 m *AOD*. The Darcy friction factor for the pipes is 0·005 and all other losses may be neglected.

Working from statements of expressions for energy loss in a pipeline and power available in a fluid stream, determine:

(*a*) the total water consumption when there is maximum power available in the water at the turbine intake; and

(*b*) the power available at the turbine intake under these conditions.

CHAPTER 3

THE ENERGY AND MOMENTUM EQUATIONS

The fundamental laws of mechanics are equally applicable to solids or to fluids. The most significant difference between solid and fluid mechanics is the fact that flowing fluids are a continuum. There are therefore corresponding differences in the way in which certain quantities are defined. The basic concepts are most easily grasped by considering how the principles of conservation of energy and conservation of momentum are applied to fluids.

CONSERVATION OF ENERGY

Energy can occur in a number of forms—potential energy, kinetic energy, strain energy, heat, to name but a few. Water in a shallow reservoir at the top of a hill possesses greater potential energy than water compounded at the bottom of the hill. If it is free to do so, the water will flow down the hill from the top, acquiring as it descends more and more kinetic energy (the energy due to motion) but losing potential energy. If in its downhill path no energy were lost in the form of heat generated by friction, the total energy of a unit mass of the water would be the same at the top and bottom of the hill, and at all points in between. In short, for each unit mass:

$$\text{Potential energy} + \text{Kinetic energy} = \text{Constant}$$

which illustrates the principle of conservation of energy.

The third form of energy in fluid flow is the energy which is derived from the action of a force or pressure. The way in which these three forms of energy interrelate becomes clear when Bernoulli's equation is considered.

BERNOULLI'S EQUATION

A stream filament is an element of fluid bounded by stream-lines. In Fig. 26, *ABCD* is a section through a stream filament, *AB* and *CD* being stream-lines. Since flow is always tangential to a stream-line, no liquid can enter or leave the stream filament except by crossing section *I* or *II*.

As a preliminary result, it is necessary to realise that if a piston of cross-sectional area a is moved a distance s against fluid in a pipe (of cross-sectional area a) at a pressure p:-

Work done on the water = Force × Distance moved by the force
= $pa \times s$
= $p \times as$
= Pressure × Volume displaced

and therefore the energy of the water is increased by an amount equal to the pressure × volume displaced.

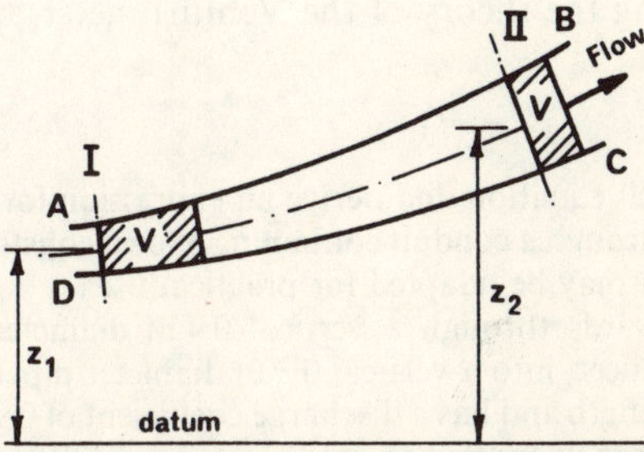

FIG. 26. Flow through a stream filament

Referring to Fig. 26, if a small volume, V, of liquid enters the stream filament at section I, a similar volume V must leave at section II, so that in flowing from I to II the liquid gains energy amounting to $p_1V - p_2V = (p_1 - p_2)V$ by virtue of the pressures p_1 and p_2 acting at I and II respectively.

Also, as the volume V flows from I to II its potential energy is increased by $\rho gV(z_2 - z_1)$ and its kinetic energy is increased by $\frac{1}{2}\rho V(u_2{}^2 - u_1{}^2)$, where u_1 and u_2 are the velocities at I and II respectively.

∴ if no energy is lost through friction,

$$(p_1 - p_2)V = \rho gV(z_2 - z_1) + \tfrac{1}{2}\rho V(u_2{}^2 - u_1{}^2)$$

This equation can be rewritten:

$$\frac{p_1}{\rho g} + \frac{u_1{}^2}{2g} + z_1 = \frac{p_2}{\rho g} + \frac{u_2{}^2}{2g} + z_2 = \text{constant, } H$$

which is one form of *Bernoulli's equation*. In this form each term has the dimensions of length (the reader should check that this is so). It follows that H also must have the dimensions of length; in fact, H is called the *total head* and is usually measured in metres.

Alternatively, each term can be thought of as expressing energy per Newton of liquid, *i.e.* N m/N.

Bernoulli's equation is a more general statement of the principle of conservation of energy and applies, strictly, only along a streamline. In practice, some energy is usually degraded into heat energy either by friction or by the formation of eddies in turbulent flow; the energy of the liquid lost in this way is denoted h_L, and Bernoulli's equation is then modified to

$$\frac{p_1}{\rho g}+\frac{u_1^2}{2g}+z_1=\frac{p_2}{\rho g}+\frac{u_2^2}{2g}+z_2+h_L$$

There follow a number of examples based on the use of Bernoulli's equation, including the theory of the Venturi meter, the Pitot tube, orifices and so on.

EXAMPLE 11

State the Bernoulli equation and derive an expression for the theoretical rate of liquid flow through a conduit containing a local constriction. Indicate how this expression may be adapted for practical use.

Water flows upwards through a vertical 0·4 m diameter pipeline and, through a taper-reducer, into a vertical 0·3 m diameter pipeline. The taper-reducer is 0·7 m in length and has a discharge coefficient of 0·98. Open-ended vertical glass tubes are connected to pressure-tappings at the extremities of the taper-reducer, and the difference between the water-levels in these tubes is observed to be 0·256 m. What is the corresponding volumetric rate of water-flow through the pipe system?

If the water pressure at the upstream end of the taper-reducer is 59 kN/m² above atmospheric pressure, what is the water pressure at the other extremity of the taper-piece?

SOLUTION

Fig. 27 (*a*) shows a pipe with a constriction at section 2–2. Denoting cross-sectional areas by a, velocities by u and pressures by p, Bernoulli's equation gives for a horizontal pipe:

$$p_1/(\rho g)+u_1^2/(2g)=p_3/(\rho g)+u_3^2/(2g)$$

where p_1 is the pressure at 1–1 and p_3 is the pressure at 3–3, etc. Also, for continuity,

$$a_1u_1=a_2u_2=a_3u_3$$

Combining these equations,

$$\frac{p_1-p_3}{\rho g}=\frac{u_3^2-u_1^2}{2g}=\frac{u_3^2}{2g}\left(1-\frac{a_3^2}{a_1^2}\right)$$

∴ the theoretical rate of flow is given by

$$a_3u_3=a_3\sqrt{2/\rho.(p_1-p_3)/(1-a_3^2/a_1^2)}$$

The liquid flowing through the constriction converges to a minimum at the *vena contracta* of area a_3, after which it diverges and becomes turbulent. Since a_2 is more conveniently measured than a_3, a *coefficient of contraction*, C_c, is defined by

$$C_c = a_3/a_2$$

$\therefore$ the theoretical rate of flow is

$$C_c a_2 \sqrt{2/\rho.(p_1 - p_3)/(1 - a_3^2/a_1^2)}$$

In practice, the constriction provides a common means of flow measurement, the actual rate of flow being given by

$$Q = C_d a_2 \sqrt{2/\rho.(p_1 - p_3)/(1 - a_3^2/a_1^2)}$$

in which C_d is the *coefficient of discharge.* A *coefficient of velocity,* C_v, is defined by $C_d = C_c C_v$ and is so called because it gives the ratio of actual to theoretical velocity at section 2–2.

Fig. 27 (*b*) is a sketch of the pipes connected by a taper-reducer.
Pressure difference causing flow = 0·256 m.

$$a_1 = \pi \times 0{\cdot}2^2\ m^2, \text{ and } a_2 = \pi \times 0{\cdot}15^2\ \text{m}^2$$

Theoretically $\quad p_1/(\rho g) + u_1^2/(2g) + z_1 = p_2/(\rho g) + u_2^2/(2g) + z_2$

But $\quad (p_1/(\rho g) + z_1) - (p_2/(\rho g) + z_2) = 0{\cdot}256\ m \qquad (1)$

$$\therefore \quad 0{\cdot}256 = (u_2^2 - u_1^2)/(2g) = u_2^2/(2g).(1 - a_2^2/a_1^2)$$
$$= u_2^2/(2g).(1 - \tfrac{81}{256})$$
$$= u_2^2/(2g).\tfrac{175}{256}$$
$$u_2 = \sqrt{0{\cdot}256 \times 2g \times 256/175} = 2{\cdot}71 \text{ m/s}$$
$$\therefore \ Q = C_d u_2 a_2 = 0{\cdot}98 \times 2{\cdot}71 \times \pi \times 0{\cdot}15^2 = \underline{\underline{0{\cdot}188 \text{ m}^3/\text{s}}}$$

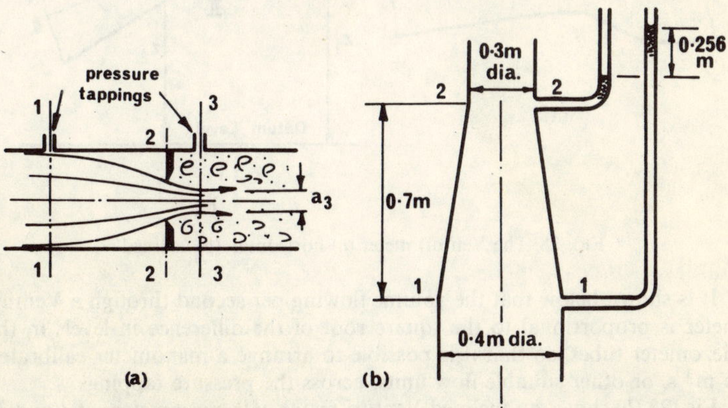

FIG. 27. *See* Example 11

From equation (1) above,

$$\frac{p_1}{\rho g} = 0{\cdot}256 + z_2 - z_1 + \frac{p}{\rho g}$$

$$\therefore \quad p_1 = [9810 \times (0{\cdot}256 + 0{\cdot}7)] + 59\,000$$
$$= 68{\cdot}4 \text{ kN/m}^2$$

Note. For a taper-reducer, and for the Venturi meter, the coefficient of discharge is defined as the ratio of actual to theoretical velocity. The same ratio for an orifice or constriction is called the coefficient of velocity; the two definitions are compatible, however, for in the taper there is no contraction and the coefficient C_c can be taken as unity.

Fig. 28 (*a*) is a sketch of a Venturi meter, a device for measuring rate of flow, which consists essentially of a taper-reducer followed by a diverging cone. In general, divergent flows break away from the boundaries, large eddies being formed which result in considerable energy losses, but this phenomenon—known as *separation*—is minimised if the angle α is less than about 10 degrees. Compared with the orifice-plate meter shown in Fig. 27 (*a*), which is also used for measuring flows, the Venturi meter is a more expensive instrument, but is more efficient in so far as greater energy losses occur downstream of the constriction of an orifice-plate meter.

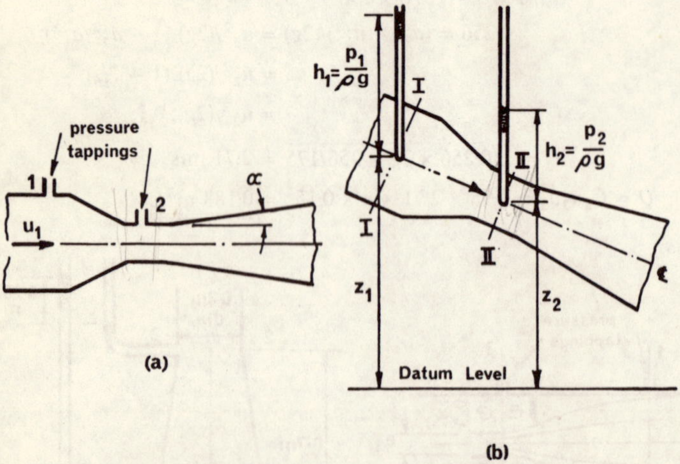

FIG. 28. The Venturi meter (*a*) horizontal, (*b*) inclined

It is shown below that the volume flowing per second through a Venturi meter is proportional to the square root of the difference in levels in the piezometer tubes, so that it is possible to arrange a manometer calibrated in m^3/s, or other suitable flow units, across the pressure tappings.

Fig. 28 (*b*) shows an inclined Venturi meter with piezometers at *I* and *II* measuring the pressures at the centre-line. Denoting conditions at these points, by the suffixes $_1$ and $_2$ respectively, and applying Bernoulli's equation,

$$(u_2^2 - u_1^2)/(2g) = (p_1 - p_2)/(\rho g) + z_1 - z_2$$

For continuity,

$$u_1 = u_2 a_2 / a_1$$

$$\therefore \frac{u_2^2}{2g}\left(1 - \frac{a_2^2}{a_1^2}\right) = \frac{p_1 - p_2}{\rho g} + z_1 - z_2$$

$$u_2^2 = \left(1 - \frac{a_2^2}{a_1^2}\right)^{-1} \times 2g\left(\frac{p_1 - p_2}{\rho g} + z_1 - z_2\right), \text{ and}$$

$$\text{Theoretical discharge} = a_2 u_2 = a_2(1 - a_2^2/a_1^2)^{-\frac{1}{2}} \sqrt{2g\left(\frac{p_1 - p_2}{\rho g} + z_1 - z_2\right)}$$

$$= a_2(1 - a_2^2/a_1^2)^{-\frac{1}{2}} \sqrt{2g(h_1 + z_1 - h_2 - z_2)}$$

When the Venturi meter is horizontal (*see* Fig. 28 (*a*)) $z_1 = z_2$ and this formula is simplified slightly. It should be noted, however, that whether the instrument is inclined or horizontal,

$$\text{Rate of flow} = \text{Constant} \times \sqrt{(\text{difference in piezometer levels})}$$

Again,

$$\text{Actual rate of flow} = C_d \times \text{Theoretical rate of flow}$$

C_d, the coefficient of discharge, usually being about 0·98 for a Venturi meter.

THE PITOT TUBE

The Pitot tube, shown diagrammatically in Fig. 29, is used for

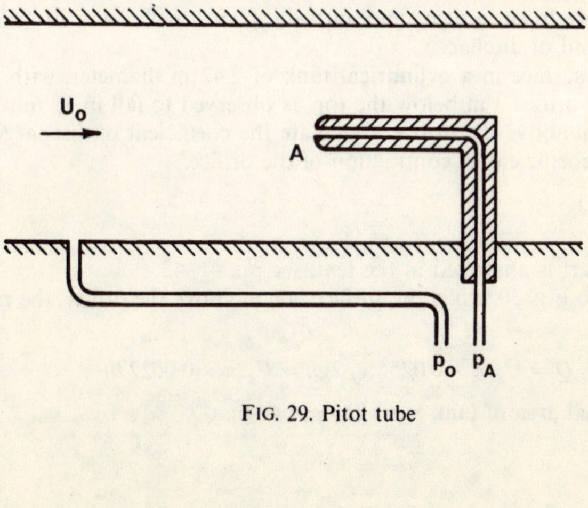

FIG. 29. Pitot tube

measuring flow velocities or local pressure intensities, particularly in wind tunnels or flow channels in which, for example, models are being tested. It must be a slender instrument so as not to interfere appreciably with the flow.

At A a stagnation point occurs; in other words, the local velocity of flow at A, u_A, is zero. Consequently Bernoulli's equation yields a relationship between p_A (the pressure at A) and the free stream velocity and pressure, u_0 and p_0 respectively, *i.e.*

$$p_0/(\rho g) + u_0^2/(2g) = p_A/(\rho g)$$

or

$$p_A - p_0 = \tfrac{1}{2}\rho u_0^2$$

FLOW THROUGH AN ORIFICE

The theoretical flow from a large tank through a relatively small hole of area a at a depth h below the free water surface can be found by the principle of conservation of energy. Assuming the tank is open to the atmosphere, the pressure at both the free surface and the orifice is atmospheric. The velocity of flow at the free surface can be taken as zero, and therefore Bernoulli's equation gives

$$h = u^2/(2g)$$

u being the theoretical discharge velocity and h being $z_1 - z_2$ in the Bernoulli equation. The actual rate of discharge is

$$Q = C_d a\sqrt{2gh}$$

C_d being as defined on p. 43. Again, $C_d = C_c C_v$.

EXAMPLE 12

State the Bernoulli equation. From it derive the formula for the flow of a liquid through an orifice, and introduce the coefficients of velocity, contraction and of discharge.

The water surface in a cylindrical tank of 2·42 m diameter, with a 50-mm-diameter orifice 3 m below the top, is observed to fall in 15 min from 2·75 to 1·25 m above the orifice. Calculate the coefficient of discharge and estimate the coefficient of contraction of the orifice.

SOLUTION

The first part is answered in the text: *see* pp. 41–43.

Referring to Fig. 30, when the surface is h m above the orifice, the rate of discharge is

$$Q = C_d.\pi \times 0{\cdot}025^2.\sqrt{2gh} = C_d.\pi \times 0{\cdot}00277h^{\frac{1}{2}}$$

Cross-sectional area of tank $= \pi 1{\cdot}21^2 = 1{\cdot}464\pi$.

Also, as the tank water-level is falling at $-\frac{dh}{dt}$ m/s,

$$Q = -1{\cdot}464\pi dh/dt$$

$$-\pi \times 1{\cdot}464 \frac{dh}{dt} = C_d\pi \times 0{\cdot}002\,77h^{\frac{1}{2}}$$

$$\therefore C_d \int_0^{15 \times 60} dt = -1{\cdot}464/0{\cdot}002\,77 \int_{1{\cdot}25}^{2{\cdot}75} h^{-\frac{1}{2}}dh$$

$$\therefore C_d = \frac{1{\cdot}464}{0{\cdot}002\,77 \times 900} \int_{1{\cdot}25}^{2{\cdot}75} h^{-\frac{1}{2}}dh$$

$$= 1{\cdot}173 \Big[h^{\frac{1}{2}} \Big]_{1{\cdot}25}^{2{\cdot}75}$$

$$\text{i.e. } C_d = 0{\cdot}635$$

The coefficient of contraction, C_c, of a sharp-edged orifice is about 0·97, and therefore,

$$C_v = C_d/C_c = 0{\cdot}635/0{\cdot}97 = 0{\cdot}655$$

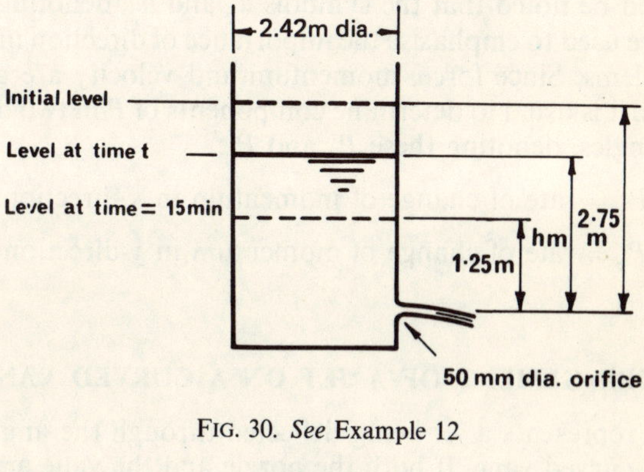

FIG. 30. *See* Example 12

CONSERVATION OF MOMENTUM

NEWTON'S second law of motion is well known in the form $P = ma$, P being the force necessary to produce an acceleration a when acting on a mass m. A more general version of the law is

Force = Rate of change of momentum

i.e. $P = d(mu)/dt$

where u is the *instantaneous* velocity.

In *steady* flow of a fluid, while the velocity may well differ from point to point in the flow, it is constant at any particular point. For example, in Fig. 26 if *ABCD* is a stream filament in a steady flow the velocity at section *I* will have the value $\bar{u}_1$ say, and that at section *II*, $\bar{u}_2$. $\bar{u}_1$ differs from $\bar{u}_2$ in both magnitude and direction, but neither $\bar{u}_1$ nor $\bar{u}_2$ changes with time. If Q is the volume of flow in m^3/s then, since there is no flow across the walls of a stream filament, the volume crossing both sections *I* and *II* is Q m^3 each second, and the mass flow across both sections is ρQ kg/s. In other words, the mass at section *I* is changing at a rate of ρQ kg/s, and consequently the momentum there is changing at the rate of $\rho Q\bar{u}_1$. Similarly, the rate of change of momentum, or *flux of momentum*, at section *II* is $\rho Q\bar{u}_2$. Evidently, steady forces must act on the fluid between *I* and *II* to produce the rate of change of momentum $\rho Q(\bar{u}_1 - \bar{u}_2)$,

i.e.
$$P = \rho Q(\bar{u}_1 - \bar{u}_2)$$

where P, the force acting on the fluid between *I* and *II*, will be the resultant of forces tangential (*i.e.* friction) and normal to the walls of the filament, the weight of the fluid and the forces due to the liquid pressure over sections *I* and *II*.

It should be noted that the symbols $\bar{u}_1$ and $\bar{u}_2$, denoting velocity *vectors*, are used to emphasise the importance of direction in momentum problems. Since force, momentum and velocity are all vector quantities, it is usual to determine components of P in two directions at right angles, denoting these P_x and P_y:

$$P_x = \text{rate of change of momentum in } x\text{-direction}$$

and
$$P_y = \text{rate of change of momentum in } y\text{-direction}$$

IMPINGEMENT OF A JET ON A CURVED VANE

Fig. 31 represents a jet being deflected through the angle β after striking a curved vane. If both the nozzle and the vane are fixed in position, then the force P acting on the vane is given by the following equations for its component forces P_x and P_y.

$$P_x = \rho Q(u_1 - u_2 \cos \beta)$$
$$P_y = \rho Q u_2 \sin \beta$$

and P is the resultant of P_x and P_y.

The forces acting on the jet are, of course, the reactions from the vane and are equal and opposite to P_x and P_y.

In the above equations $Q = a_1u_1 = a_2u_2$, where a_1 and a_2 are the cross-sectional areas at sections *I* and *II* respectively.

If the vane were not fixed but moving in the direction of P_x with a velocity v, then the above equations would have to be modified. In this case the mass flowing on to the vane in 1 second is no longer ρQ kg but $\rho a_1(u_1 - v)$ kg; u_2 in Fig. 31 would be the exit velocity *relative to the vane* instead of the actual exit velocity; and finally,

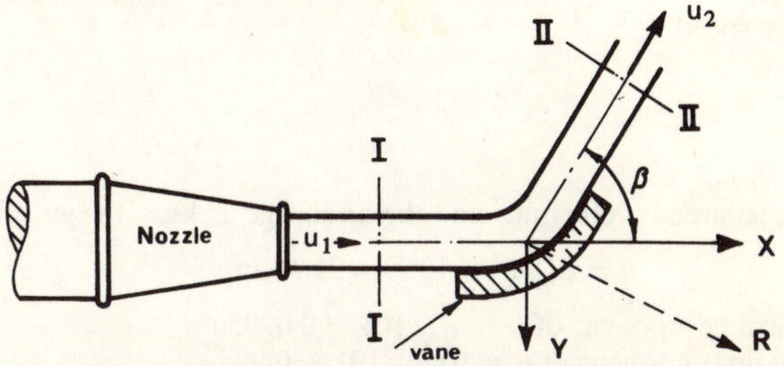

FIG. 31. Jet deflected by a curved vane

the approach velocity u_1 must be replaced by $(u_1 - v)$. These modifications give an example of the "bringing to rest" technique described below.

"BRINGING TO REST" TECHNIQUE

There is no way of determining absolute velocity. For example, we say a car is travelling at 50 km/h when we mean that its velocity *relative to the earth* is 50 km/h. The velocity of the earth relative to the sun can be determined and, to some extent, the velocity of the sun relative to the stars can be estimated, but these relationships do not help in determining the absolute velocity of the car, for there is no single point in the universe which we know to be truly at rest. In fact, in this context the phrase "at rest" is meaningless. We are conscious of velocity only in so far as it is relative to something else, frequently the earth.

In the same way, momentum is a relative quantity. We are conscious of rate of change of momentum (*i.e.* force), but not of momentum itself. Consequently, in any one problem, it does not matter from what datum we measure momentum, providing the same datum is used throughout the problem. For example, if we wish to determine the force with which a jet impinges on a moving vane, we can measure all the velocities and rates of flow *relative to the vane* and then apply the momentum principle. This procedure may be described as bringing the vane to rest, for relative to itself it has zero velocity. This technique is employed in the following example.

EXAMPLE 13

An experimental jet-propelled barge is under test in a fresh-water canal. The drag D of the barge, found to be proportional to the square of the speed v, is $D = 45$ N at $v = 1$ m/s. At full speed the submerged outlet (40 mm diameter) of the jet pipe is inclined at 1 in 10 downward. Water flows through the jet pipe at the rate of 27 kg/s. Calculate the horizontal and vertical components of the jet thrust when the barge is: (*a*) at rest, and (*b*) at its steady speed.

SOLUTION

See Fig. 32.

Since the jet area is 400π mm^2 and the discharge 27 kg/s, the jet velocity is

$$u = 27/10^3 \times 10^4\pi = 21{\cdot}45 \text{ m/s}$$

(*a*) Vertical component of $u = u/\sqrt{101} = 0{\cdot}0995u$
Horizontal component of $u = 10u/\sqrt{101} = 0{\cdot}995u$.

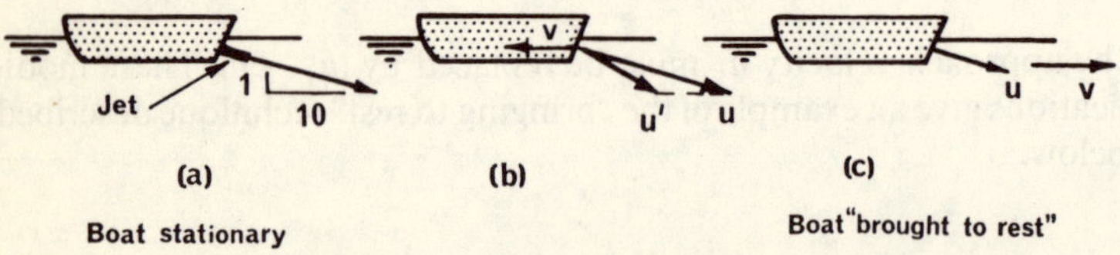

FIG. 32. (*a*) *See* part (*a*) of Example 13. (*b*) and (*c*) *See* part (*b*) of Example 13. (*b*) shows the velocity, v, of the boat relative to the water, and the velocity of the jet relative to the water, u'. (*c*) shows the boat "brought to rest" so that the water is now flowing past it with velocity v

The jet is quickly dissipated, and at some distance behind the boat its velocity is zero.

∴ Horizontal component of force

$$= \text{Rate of change of horizontal momentum}$$
$$= 27 \times 21{\cdot}45 \times 0{\cdot}995 = \underline{\underline{577{\cdot}0 \text{ N}}}$$

Similarly, vertical component of force

$$= 27 \times 21{\cdot}45 \times 0{\cdot}0995 = \underline{\underline{57{\cdot}7 \text{ N}}}$$

(*b*) $$D = \text{constant} \times v^2$$

and since $D = 45$ when $v = 1$, the constant $= 45$, and $D = 45v^2$. The barge will attain its steady speed when the drag equals the horizontal thrust, *i.e.* when

$$D = \rho Q(0{\cdot}995u - v)$$
i.e. when $45v^2 = 27(0{\cdot}995 \times 21{\cdot}45 - v)$
i.e. when $v = 3{\cdot}28$ m/s

At this speed, horizontal thrust $= 27(21{\cdot}45 \times 0{\cdot}995 - 3{\cdot}28)$
$= \underline{\underline{488\ \text{N}}}$

As before, vertical thrust $= \underline{\underline{57{\cdot}7\ \text{N}}}$

QUESTIONS

1. The rate of flow of air in a circular pipe 300 mm diameter is measured by means of a Venturi meter of throat diameter 150 mm. The Venturi meter is connected to an inclined manometer as shown in Fig. 33. The manometer liquid is kerosene (relative density 0·8) and the surface area in the container is fifty times that in the inclined tube.

Determine the rate of flow of air when the manometer scale reading S is 200 mm. Assume an air density of 1·3 kg/m^3 and ignore compressibility effects.

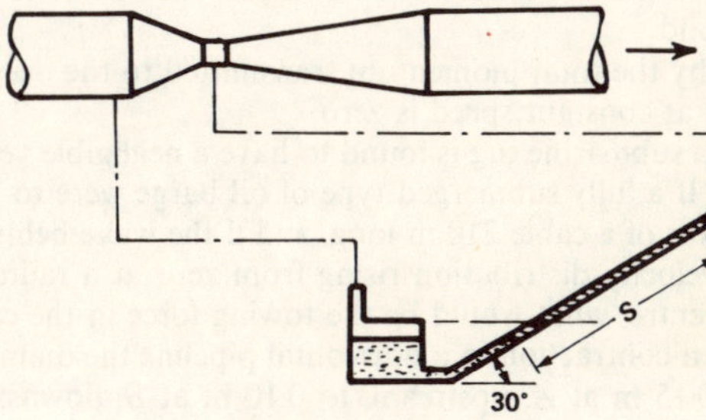

FIG. 33. *See* Question 1

2. A cylindrical water tank, 1·5 m in diameter and 3·4 m high, stands with its axis vertical and its top open. Two sharp-edged orifices are located in the bottom of the tank and discharge water freely into a collecting channel underneath. Each orifice has a diameter of 75 mm and a discharge coefficient of 0·62. Determine, from first principles, the rate at which water must be fed into the tank in order to maintain the water surface at an elevation 2 m above the tank bottom.

If, under these conditions, one orifice suddenly becomes totally blocked, determine the time interval before the tank overflows.

3. State the Bernoulli equation and from it derive an expression for the theoretical rate of flow of an incompressible fluid through a conduit containing a local constriction. Indicate how the expression may be modified for practical use.

A water flow of up to 35 litres/s is expected to pass through a 120 mm internal-diameter pipeline and is to be measured by means of a Venturi meter having the same upstream diameter as the pipeline and a throat diameter of 55 mm. If the discharge coefficient for the meter is assumed to be 0·97 what maximum deflection will be indicated by a suitably connected mercury and water U-tube manometer?

Discuss briefly the relative merits of Venturi meters and orifice-plate meters for measuring the rates of water flow in a pipeline.

4. Define the coefficient of velocity, contraction and discharge of an orifice.

Calculate all three coefficients for an orifice of 25 mm diameter set into the side of a large open tank, at 2·44 m below the water surface. The orifice discharge has been found to be 2·0 kg/s and the centre-line of the jet passes through a point whose horizontal and vertical distances from the vena contracta are 2·24 m and 0·6 m respectively.

[*Hint:* The text associated with Fig. 35 (*b*) may be helpful in solving this problem.]

5. Water is pumped along a pipe of 76 mm diameter. A horizontal part of the pipe contains a Venturi throat of 38 mm diameter. The loss of head in the Venturi between the manometer points is equivalent to a 1 m head of water.

The difference of head between the throat and the upstream, parallel portion of the pipe is measured by a mercury manometer, the communicating tubes between pipe and manometer being full of water. Calculate the difference of levels in the two limbs of the U-tubes when 14 litres of water are passing per second.

6. Explain why the total momentum transmitted to the ocean by a submarine cruising at constant speed is zero.

The wake of a submarine tug is found to have a negligible velocity 180 m behind the tug. If a fully submerged type of oil barge were to be towed by that tug by means of a cable 210 m long, and if the wake behind the barge had a conical velocity distribution rising from zero at a radius of 3 m to 1·5 m/s at the centre, what would be the towing force in the cable?

7. At a sudden contraction in a horizontal pipeline the diameter changes abruptly from 0·15 m at *A*, upstream, to 0·10 m at *B*, downstream of the contraction. When the flow of water is 30 litres/s the gauge pressure at *A* is 40 kN/m^2 and the energy loss due to the contraction is estimated to be 0·38 of the kinetic energy at *B*.

Determine the force that will be exerted on the annular disc at the change in section.

CHAPTER 4

APPLICATIONS OF THE MOMENTUM AND ENERGY EQUATIONS

IN Chapter 3 it was seen that Force = Rate of change of momentum. The left-hand side of this equation may be the resultant of a number of contributory forces, among which there are often forces due to fluid pressure. The relationship between fluid pressures is given by Bernoulli's equation (Chapter 3), and consequently, in many flow problems, it is necessary to apply the principle of *conservation of energy* before determining force by the principle of *conservation of momentum.* Solutions to a selection of such problems are given in this chapter.

PIPE BENDS, TAPERS AND NOZZLES

Referring to Fig. 34, subscript $_1$ refers to section *I–I* and subscript $_2$ to section *II–II*. Velocities are denoted by u, pressures by p, areas by a and diameters by d,

(a) In applying the momentum principle to water flowing through pipes it is usually more convenient to consider the force on the water due to the pipe rather than the reverse (*i.e.* the force on the pipe due to the water). Thus, in Fig. 34 (*b*) F represents the force due to the shape of the pipe shown in Fig. 34 (*a*) on the water flowing through it. F_x and F_y are the components of F.

(*b*) For continuity, $u_1a_1 = u_2a_2 = Q$

$$\therefore u_1/u_2 = a_2/a_1 = d_2^{\,2}/d_1^{\,2}$$

(*c*) The friction loss in a *short length* of pipe is usually negligible, so that by Bernoulli's equation,

$$\frac{p_1}{\rho g} + \frac{u_1^{\,2}}{2g} + z_1 = \frac{p_2}{\rho g} + \frac{u_2^{\,2}}{2g} + z_2$$

If the pipe lies in a horizontal plane, $z_1 = z_2$, and the Bernoulli equation is simplified by the omission of these terms.

(*d*) The Bernoulli equation only gives the relationship between p_1 and p_2. The actual value of either can be found only if the value of the other is known. In Fig. 34 (*c*), for example, the value of p_2 is the same as the pressure in the medium surrounding the jet—usually atmospheric pressure. Were this not so, the jet would contract or expand, for the water in it would have a radial component of velocity owing to the pressure gradient that would exist in the radial direction.

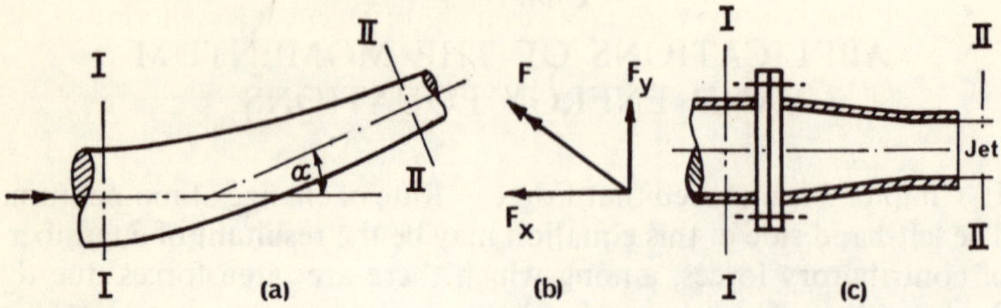

FIG. 34. Tapered pipe bend and nozzle. (*a*) Tapered pipe bend. (*b*) Showing the directions of the components of the force due to the pipe on the water in (*a*). (*c*) Section through a nozzle

(*e*) In Fig. 34 (*a*) the momentum principle yields the equations:

$$p_1 a_1 - F_x - p_2 a_2 \cos\alpha = \rho Q(u_2 \cos\alpha - u_1)$$

and
$$F_y - p_2 a_2 \sin\alpha = \rho Q u_2 \sin\alpha$$

(*f*) In Fig. 34 (*c*) the force F due to the nozzle on the water flowing through it is given by

$$p_1 a_1 - F - p_2 a_2 = \rho Q(u_2 - u_1)$$

Providing p_1 and p_2 are *gauge pressures*, the value of F derived from this equation is also the force on the bolts connecting the flanges (assuming the metal wall of the nozzle is of negligible thickness). If p_1 and p_2 are not gauge pressures it is necessary, in order to calculate the force on these bolts, to include a term giving the force due to atmospheric pressure on the outside of the nozzle.

EXAMPLE 14

A fluid-flow is described as uniform, unsteady and turbulent. What does this mean?

A water-jet, with mean velocity of 22 m/s and diameter 25 mm, issues from a nozzle at the end of a 65 mm diameter fire-hose. The walls of the nozzle are parallel at outlet, and the outlet diameter is 25 mm. If the frictional losses within the nozzle are small, determine the magnitude and direction of the resultant force on the nozzle caused by the fluid-motion. Also, calculate the reading which will be shown by a Bourdon pressure-gauge directly connected to a pressure-tapping at the point of connection of the nozzle and fire-hose.

Plot a graph of jet-velocity (m/s) against pressure-gauge reading (m of water) and suggest a convenient method for verifying the relationship by experiment.

SOLUTION

First part.—These terms are discussed in Example 8. An example of uniform, unsteady, turbulent flow is a water column being accelerated by a piston in a pipe. This flow would be unsteady by virtue of its acceleration and, providing the velocity were great enough, turbulent but uniform (in the broad sense).

Second part.—Fig. 34 (*c*) is a suitable sketch of the hose and nozzle if we take $p_2 = 0\ \text{N/m}^2$, $d_1 = 65$ mm diameter and $d_2 = 25$ mm, and $u_2 = 22$ m/s.

For continuity, $\frac{1}{4}\pi {d_1}^2 u_1 = \frac{1}{4}\pi {d_2}^2 u_2$

$$\therefore {u_1}^2 = 22^2 \times (25/65)^4 = 22^2 \times 1/45{\cdot}7$$

Bernoulli's equation states:

$$p_1/(\rho g) - p_2/(\rho g) = {u_2}^2/(2g) - {u_1}^2/(2g) \qquad (1)$$

$\therefore$ since $p_2 = 0\ \text{N/m}^2$ gauge,

$$p_1 = \tfrac{1}{2}\rho({u_2}^2 - {u_1}^2) = \tfrac{1}{2} \times 10^3 \times 22^2(1 - 1/45{\cdot}7)\ \text{N/m}^2$$

$$\therefore p_1 = 236{\cdot}7\ \text{kN/m}^2 \text{ gauge pressure}$$

$$\text{Rate of flow } Q = 22 \times \frac{\pi}{4} \times 0{\cdot}025^2 = 0{\cdot}0108\ \text{m}^3/\text{s}$$

If F is the force exerted on the water by the nozzle the momentum equation (*see* note (*f*) above) gives:

$$236{\cdot}7 \times 10^3 \times \tfrac{1}{4}\pi \times 0{\cdot}065^2 - F = 10^3 \times 0{\cdot}0108 \times 22(1 - 25^2/65^2)$$

from which

$$\underline{\underline{F = 583\ \text{N}}}$$

Since a Bourdon pressure-gauge records gauge pressure, its reading will be p_1, *i.e.*

$$\underline{\underline{236{\cdot}7\ \text{kN/m}^2}}$$

From equation (1), the pressure-gauge reading in m of water is

$$p_1/(\rho g) = {u_2}^2/(2g) \times (1 - 1/45{\cdot}7) = 0{\cdot}0498{u_2}^2 \text{ m water}$$

which when plotted gives the graph shown in Fig. 35 (*a*).

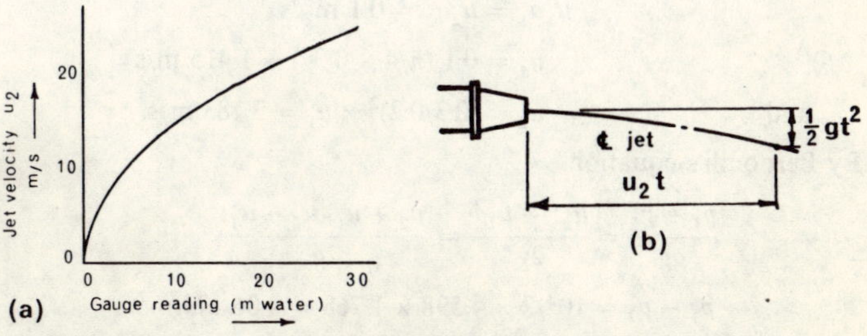

FIG. 35. *See* Example 14

One method of verifying the result experimentally is to calculate the actual jet velocity from the parabolic properties of the jet. If the jet issues horizontally from the nozzle with velocity u_2 it will, in time t, travel a horizontal distance u_2t and fall a vertical distance $\frac{1}{2}gt^2$ (*see* Fig. 35 (*b*)). The value of u_2 can therefore be obtained by measuring these co-ordinates, while the corresponding values of pressure can be read from a manometer.

EXAMPLE 15

A converging pipe-bend, with its centre-line in a horizontal plane, changes the direction of a pipeline by 60 degrees and reduces the pipeline diameter from 0·30 m to 0·20 m in the direction of flow. If the pressure indicated by a Bourdon gauge at the centre-line of the 0·30 m diameter entrance to the bend is 140 kN/m², and the flow of water through the pipeline is 0·10 m³/s, determine the magnitude and direction of the force on the bend due to the moving water.

SOLUTION

Fig. 36 (*a*) is a sketch of the pipe-bend and Fig. 36 (*b*) shows the resultant force R due to the pipe-bend on the water, F_x and F_y being the components of R. Using the notation described on p. 53.

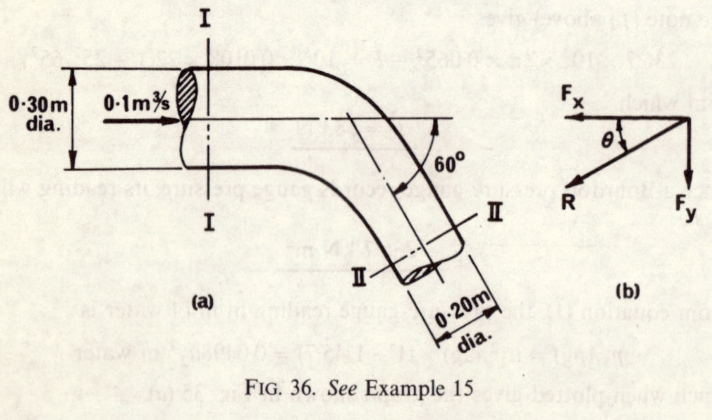

FIG. 36. *See* Example 15

$$u_1a_1 = u_2a_2 = 0{\cdot}1 \text{ m}^3/\text{s}$$

$$u_1 = 0{\cdot}1/(\pi/4 \times 0{\cdot}3^2) = 1{\cdot}415 \text{ m/s}$$

and $$u_2 = (0{\cdot}3/0{\cdot}2)^2 \times u_1 = 3{\cdot}183 \text{ m/s}$$

By Bernoulli's equation,

$$\frac{p_1 - p_2}{\rho g} = \frac{u_2{}^2 - u_1{}^2}{2g} = \frac{(u_2 + u_1)(u_2 - u_1)}{2g}$$

$$p_1 - p_2 = 10^3/2 \times 4{\cdot}598 \times 1{\cdot}768 = 4{\cdot}06 \text{ kN/m}^2$$

Since $p_1 = 140$ kN/m², $p_2 = 135{\cdot}9$ kN/m² (gauge).

Applying the momentum principle, first in the x-direction and then in the y-direction:

$$140 \times 10^3 \times \pi \times (0{\cdot}15)^2 - F_x - 135{\cdot}9 \times 10^3 \times \pi \times (0{\cdot}10)^2 \cos 60° = 10^3 \times 0{\cdot}1(3{\cdot}183 \cos 60° - 1{\cdot}415)$$

from which $F_x = 7{\cdot}744$ kN

$$F_y - 135{\cdot}9 \times 10^3 \times \pi \times (0{\cdot}10)^2 \sin 60° = 10^3 + 0{\cdot}1 \times 3{\cdot}183 \sin 60°$$

from which $F_y = 3{\cdot}973$ kN

Referring to Fig. 36 (*b*),

$$R = \sqrt{7{\cdot}744^2 + 3{\cdot}973^2} = \underline{\underline{8{\cdot}70 \text{ kN}}}$$

and $\theta = \tan^{-1}(3{\cdot}973/7{\cdot}744) = \underline{\underline{27{\cdot}2°}}$

The force on the bend due to the moving water is, of course, equal and opposite to R.

RAPID CLOSURE OF VALVES

From Bernoulli's equation it is clear that, if the friction losses are neglected, the total energy of the water flowing in a pipe is the same at all points along the pipe. If the water is suddenly brought to rest by closing a valve this energy does not simply disappear, but, in fact, is converted into strain energy. Although it may be convenient to assume that it takes no time at all to close the valve (we then say that the valve closure is "sudden"), it is wrong to assume that all the water is brought to rest in an instant. What happens is shown diagrammatically in Fig. 37 (*a*) and is described below.

Immediately the valve is closed a shock wave *SS* is formed which, starting at the valve, travels against the direction of flow at a speed c, called its propagation velocity. The wave separates the water flowing at speed u, which is at relatively low pressure, from the water at rest, which is under very high pressure, the deceleration taking place in the space of the shock wave. An everyday analogy is the interruption of a steady flow of traffic along a main road by traffic lights turning red: the cars do not all stop at the same instant, those nearest the lights being at rest, while others well away from the lights are still travelling at their steady speed. An observer from a high building would see a shock wave moving against the flow, and separating the stationary cars near the lights from those still moving. The density of traffic (corresponding to water pressure) would be high where the cars are stationary, and closely parked, and low where the traffic is still moving.

Referring to Fig. 37 (*a*), for conservation of energy, Loss of kinetic energy (K.E.) of the water = Gain of strain energy (S.E.).

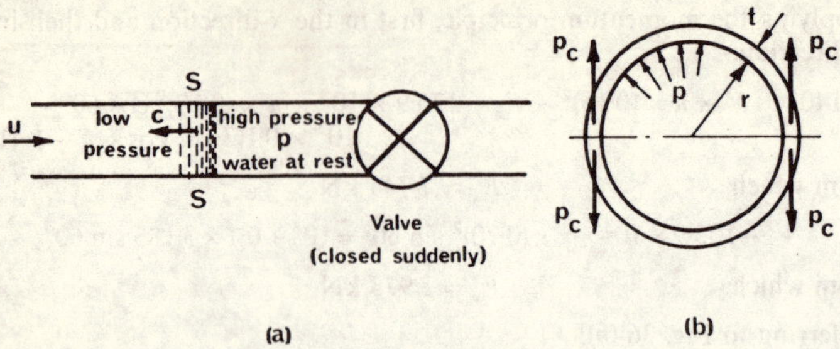

FIG. 37. Water hammer

$$\text{Loss of K.E. per m of pipe} = \tfrac{1}{2}\rho\pi r^2 u^2$$

where r is the radius of the pipe.

From the definition* of K, the bulk modulus of elasticity,

$$\text{Volumetric strain of the water in the pipe} = p/K$$

Hence,

$$\text{S.E. of water per m of pipe} = \tfrac{1}{2}\ \text{stress} \times \text{Strain} \times \text{Volume}$$
$$= \tfrac{1}{2}p \times p/K \times \pi r^2$$

Referring to Fig. 37 (*b*), if E is Young's modulus for the pipe,

$$\text{S.E. of pipe per m of pipe} = \tfrac{1}{2}p_c \times p_c/E \times \text{Volume}$$
$$= \tfrac{1}{2}p_c{}^2/E \times 2\pi rt$$
$$= \tfrac{1}{2}p^2/E \times 2\pi r^3/t$$

because $2rp = 2tp_c$.

Hence, $\tfrac{1}{2}\rho\pi r^2 u^2 = \tfrac{1}{2}p^2\pi r^2[1/K + 2r/(tE)]$

$$\therefore\ p = u\sqrt{\frac{\rho}{\dfrac{1}{K} + \dfrac{2r}{tE}}} \qquad (1)$$

If the pipe is *rigid*, which means it is not free to expand radially, this formula becomes

$$p = u\sqrt{\rho K} \qquad (2)$$

* A volume V of fluid compressed by an all-round pressure p suffers a decrease in volume, dV, given by

$$p = -K.\mathrm{d}V/V$$

where K is the *bulk modulus of elasticity*, which, like other moduli of elasticity, may be thought of as stress/strain, *i.e.* $p/(\mathrm{d}V/V)$. It is also called the *coefficient of compressibility*.

FIG. 38. Velocities of flow relative to the shock wave *SS*

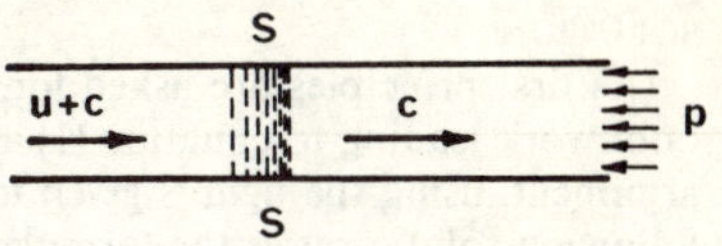

Propagation velocity

Fig. 38 shows the shock wave brought to rest so that the momentum principle can more easily be applied to the flow across it. The pressure upstream of the shock wave is taken as zero, for it is negligible compared with the high pressure p downstream of SS. The product of p and the cross-sectional area a of the pipe is therefore, the only force producing a change of momentum. Hence,

$$pa = (u + c)a\rho\{(u + c) - c\}$$

$$\therefore\ p = \rho u(u + c) \simeq \rho uc, \qquad \text{since } u \ll c$$

Hence, the propagation velocity,

$$c = p/(\rho u) \tag{3}$$

From equation (1) above,

$$c = \sqrt{\frac{1}{\rho\left(\dfrac{1}{K} + \dfrac{2r}{tE}\right)}} \tag{4}$$

Note that, if the pipe is rigid, equation (2) applies and the propagation velocity becomes

$$c = \sqrt{K/\rho} \tag{5}$$

which is the same as the velocity of sound in a fluid of density ρ.

EXAMPLE 16

A steel pipeline has a length of 14·5 km, an internal diameter of 0·9 m and a wall thickness of 16 mm. The pipeline flows full and carries water at a mean speed of 2 m/s. If a valve at the downstream end of the pipeline is closed quickly, and if it can be assumed that the pipeline cannot move in a longitudinal direction, show, using first principles, how the pressure rise at the valve can be predicted and obtain a value for this pressure rise.

What time interval would be expected between successive pressure rises at the valve? It may be assumed that:

for steel, $E = 200\ \text{GN/m}^2$

for water, $K = 2\ \text{GN/m}^2$

SOLUTION

(As first principles are asked for, it is necessary either to reproduce the bookwork leading to equation (1) and (3), or to follow the same theoretical argument, using the figures given in the question in place of symbols. The following solution uses the formulae already derived.)

$$\frac{1}{K}+\frac{2r}{tE}=\frac{1}{2\times 10^9}+\frac{2\times 0{\cdot}45}{0{\cdot}016\times 2\times 10^{11}}$$

$$=\frac{1}{2\times 10^9}\left(1+\frac{0{\cdot}9}{1{\cdot}6}\right)=\frac{2{\cdot}5}{3{\cdot}2\times 10^9}$$

$$\therefore p=u\sqrt{\frac{\rho}{\dfrac{1}{K}+\dfrac{2r}{tE}}}=2\times\sqrt{\frac{3{\cdot}2\times 10^9\times 10^3}{2{\cdot}5}}\ \text{N/m}^2$$

$$\therefore \frac{p}{\rho g}=\frac{2\times 10^6}{9810}\sqrt{\frac{3{\cdot}2}{2{\cdot}5}}\ \text{m of water}$$

$$=\underline{\underline{230{\cdot}7\ \text{m}}}$$

From equation (3), $c=p/(\rho u)=\dfrac{2\times 10^6}{10^3\times 2}\sqrt{\dfrac{3{\cdot}2}{2{\cdot}5}}=1131$ m/s

Since the wave has to travel the length of the pipe and back again twice (*see* next section) between successive pressure rises at the valve, the time interval between these is:

$$\frac{4\times 14{\cdot}5\times 10^3}{1131}=\underline{\underline{51{\cdot}3\ \text{seconds}}}$$

Water hammer

In a pipeline of length L the pressure wave SS will travel the whole length of the pipe in time L/c, *i.e.* if at time $t=0$, SS is at the valve, then at time $t=L/c$, SS reaches the other end of the pipe. After time $t=L/c$, therefore, the pipe is full of stationary high-pressure liquid, the adverse pressure gradient now causing a reverse flow from the pipe into the low-pressure liquid (in a reservoir, for example) beyond the open end of the pipe. The water in the pipe is accelerated to the reverse flow velocity across the reflected shock wave which, consequently, reaches the valve after a further time interval of L/c; *i.e.* at time $t=2L/c$, the pipe is full of water moving with velocity $-u$.

During the next time interval of L/c the water is again brought to rest by the shock wave moving from the valve to the open end, leaving the pipe full of very low-pressure water. The cycle is completed in the fourth interval of L/c, during which the water in the pipe is accelerated to its original velocity u towards the valve.

Summarising, high pressures occur at the valve at times $t = 0$, $4L/c$, $8L/c$ ... and low pressures occur there at times $t = 2L/c$, $6L/c$, $10L/c$ ... The fluctuating pressure accounts for the familiar knocking heard, for example, after turning off the domestic water supply too rapidly. This phenomenon is sometimes known as *water hammer*.

In practice, the extreme pressures are gradually attenuated as the strain energy is degraded into heat energy, so that the knocking dies away.

Throughout, it has been assumed that the valve is closed in an instant. Since the high pressure at the valve exists from time $t = 0$ to $t = 2L/c$, it follows that water hammer will occur (with the pressure as given by equations (1) or (2)), providing the time taken to close the valve is less than $2L/c$.

SLOW CLOSURE OF VALVES

Valve closure is described as *slow* when the water is brought to rest without a shock wave being formed. The subtlety of this condition is illustrated in Example 17, which shows how a weakened shock wave can be propagated, even though the time to close the valve gently exceeds the value $2L/c$.

In problems where the valve is said to be closed *slowly* it must be assumed that a shock wave does not form. A further assumption, that all the water in the pipeline is brought to rest with uniform deceleration, is also made. Then, by a straightforward application of Newton's second law, the force retarding the flow is given by:

Force = Mass of water in pipe × Its uniform deceleration

The resulting pressure rise is given by dividing the force by the cross-sectional area of the pipe—as is done in the first part of the solution to Example 17.

EXAMPLE 17

In a hydro-electric scheme with a gross head of 450 m, the 900 m long pipeline connecting the large surge shaft to the spear valve at the turbine is embedded in concrete and rock of great rigidity. The speed of flow may reach 9 m/s, and on complete rejection of full load the valve is designed to close at a uniform rate in 45 seconds. Calculate the resultant total peak pressure at the downstream end of the pipeline.

The analysis of the valve movement shows that it would close completely in 1 second from the position in which the speed in the pipeline is 0·6 m/s. Estimate the resulting water-hammer pressure.

The bulk modulus of elasticity of water is 2·0 GN/m^2.

SOLUTION

$$\text{Uniform deceleration} = \frac{9}{45} = 0{\cdot}2\ \text{m/s}^2$$

If the cross-sectional area of the pipe is a the pressure-rise produced by this deceleration is given by:

$$pa = 900a\rho \times 0{\cdot}2\ \text{N}$$

i.e.

$$p = 180 \times 10^3 = 180\ \text{kN/m}^2$$

The gross hydrostatic head is 450 m $= 450 \times 9{\cdot}81 \times 10^3$

$$= 4420\ \text{kN/m}^2$$

$$\therefore \text{Total downstream pressure} = 180 + 4420$$

$$= \underline{\underline{4{\cdot}6\ \text{MN/m}^2}}$$

From equation (2) on p. 58,

$$\text{water-hammer pressure} = u\sqrt{\rho K}$$

$$= 0{\cdot}6\sqrt{10^3 \times 2{\cdot}0 \times 10^9}$$

$$= \underline{\underline{0{\cdot}849\ \text{MN/m}^2}}$$

Notice that the water-hammer pressure of 0·849 MN/m^2 is very much greater than the pressure rise of 180 kN/m^2 calculated by applying the momentum principle.

SURGE TANKS

In hydro-electric installations the turbines are driven by water which has to be brought from the reservoir by long pipes or tunnels, often of large cross-sectional area. The turbine speed is controlled by an automatic governing mechanism which, if it shuts too quickly, can produce a deceleration of the water supply which approaches the rapid valve closure conditions described above with the danger of very high, possibly destructive, pressures being developed. Clearly, a safety device is necessary and is usually provided in the form of a *surge tank* built near the turbines so that in the event of rapid shut-down the approaching water in the supply pipeline has an escape route. It can flow into the surge tank, raising the level there and dissipating the pressure build-up. The pressure in the pipeline can be limited to a pre-arranged maximum value by leaving the surge tank open at the top at a suitably modest height above the reservoir water level, so that in extreme conditions water spills over.

A second function of a surge tank is to provide a readily available head when the turbines are starting up and the water in the supply pipe is accelerating to meet the demand. It is during this starting-up period that the water level in the surge tank will fall to its lowest level. Under steady operating conditions its water level will be below that in the reservoir by an amount equal to the friction head in the supply pipe.

A surge tank has to be of adequate strength and size and yet built at minimum cost. Other design considerations are the period and amplitude of the oscillations of its water level following shut-down of the turbines. These oscillations are the subject of Example 18 which deals with the case of frictionless flow only. The inclusion of friction complicates the mathematics which then can be solved only by a method of successive approximations.

EXAMPLE 18

A surge shaft as shown in Fig. 39 is supplied by two pressure tunnels from two reservoirs. Neglecting friction, determine the period of oscillation of the surge for total sudden rejection of load. Work from first principles.

Also determine the maximum rise of water level in the shaft.

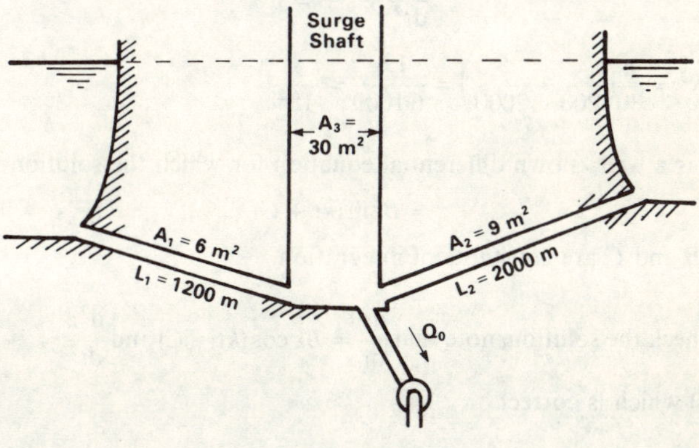

FIG. 39. *See* Example 18

SOLUTION

Applying Newton's second law of motion to the water in the left-hand tunnel (and measuring z from the bottom of the shaft):

$$\text{force} = \text{mass} \times \text{acceleration}$$

i.e.
$$A_1 \times \rho g z = -\rho A_1 L_1 \times \frac{du_1}{dt}$$

$\therefore$
$$z = -\frac{L_1}{g}\frac{du_1}{dt}$$

where u_1 is the velocity in the left-hand pipe. Since $L_1 = 1200$

$$z = -\frac{1200}{g}\frac{du_1}{dt}$$

Similarly, for the right-hand pipe (velocity = u_2)

$$z = -\frac{2000}{g}\frac{du_2}{dt}$$

For continuity, the flow in the shaft = sum of the flows in the pipes,

i.e. $$A_3u_3 = A_1u_1 + A_2u_2$$

where u_3 = flow velocity in the shaft $= \frac{dz}{dt}$. Putting in the values of A_1, A_2, A_3, the continuity equation becomes:

$$30\frac{dz}{dt} = 6u_1 + 9u_2$$

$$\therefore \quad 30\frac{d^2z}{dt^2} = 6\frac{du}{dt} + 9\frac{du}{dt}$$

$$= -\frac{g}{200}z - \frac{9g}{2000}z$$

i.e. $$\frac{d^2z}{dt^2} + k^2z = 0$$

where $k^2, = \frac{g}{30}\left(\frac{1}{200} + \frac{9}{2000}\right) = \frac{19g}{60\,000} \simeq \frac{1}{18^2}$

This is a well known differential equation for which the solution is:

$$z = B\sin(kt + C)$$

where B and C are constants of integration.

To check the solution, note that $\frac{dz}{dt} = Bk\cos(kt + C)$ and $\frac{d^2z}{dt^2} = -Bk^2\sin(kt + C)$ which is correct.

To evaluate C: when $t = 0$, $z = 0$ and, therefore, $C = 0$.

To evaluate B: when $t = 0$, $\frac{dz}{dt} = \frac{Q_0}{30k} = Bk$ and, therefore, $B = \frac{Q_0}{30k}$.

Hence,

$$z = \frac{Q_0}{30k}\sin kt$$

i.e. $$z = 0{\cdot}6Q_0\sin\frac{t}{18} \qquad \text{since } k = 1/18$$

The oscillations are sinusoidal and therefore are repeated each time the angle increases by 2π, *i.e.* when $t/18 = 2\pi$. Hence:

period = 36π and amplitude = $0{\cdot}6Q_0$

QUESTIONS

1. A horizontal pipe, 0·2 m in diameter and 100 m long discharges water into atmosphere from a reservoir. If the operating head is 11 m and the Darcy friction factor for the pipe is 0·01, calculate the maximum pressure in the pipe when a valve is suddenly closed at its outlet. Assume the walls of the pipe are rigid and that the Bulk modulus of water = 2×10^9 N/m^2.

2. A steel pipe 2000 m long, of 150 mm internal diameter and 10 mm wall thickness, carries a steady discharge of 0·05 m^3/s of water. Calculate the maximum hoop stress in the pipe due to rapid closure of a valve at the downstream end and the greatest duration of the valve closing movement producing this pressure. The bulk modulus of water is $K = 2 \times 10^6$ kN/mm^2 and Young's modulus for steel $E = 200$ kN/mm^2.

3. (*a*) A pipe of internal diameter D and wall thickness t carries a steady flow of liquid at a mean velocity $\bar{u}$. The density of the liquid is ρ and its bulk modulus of compressibility is K.

Show that, for a pipe which is not restrained longitudinally, a value for the pressure rise Δp in the liquid when a valve is suddenly closed can be obtained from:

$$\Delta p = \bar{u}\left[\frac{1}{\rho}\left\{\frac{1}{K} + \frac{D}{tE}\right\}\right]^{-1/2}$$

where E is the modulus of elasticity of the material of the pipe.

(*b*) A cast-iron pipe is 0·15 m bore and 15 mm thick. It carries a steady flow of water. Using the above formula, determine the maximum value of the flow rate in m^3/s if, on sudden closure of a valve, the increase in the hoop stress in the pipe wall is not to exceed 50,000 kN/m^2. For water, $K = 2 \times 10^6$ kN/m^2 and, for cast-iron, $E = 1{\cdot}24 \times 10^8$ kN/m^2.

4. A 50 mm diameter faired fire nozzle discharges a cylindrical jet of water at a velocity of 24 m/s. Calculate the hydraulic force on the bolts connecting the nozzle to the flange of the 90 mm diameter hose.

5. A pipe of 80 mm diameter terminates in a nozzle which delivers a smooth horizontal jet of 20 mm diameter. The pressure in the pipe at the coupling to the nozzle is 875 kN/m^2 when the rate of discharge is 12·5 litres/second. Determine:

(*a*) the head loss in the nozzle;
(*b*) the nozzle velocity coefficient;
(*c*) the power available in the jet;
(*d*) the force exerted by the jet on a flat plate placed perpendicularly to the jet a short distance from the nozzle.

6. Discuss briefly the problems associated with the governing of water-turbines and suggest ways of surmounting these problems.

Water flows at a mean speed of 1·7 m/s through a steel pipeline which has an internal diameter of 1·3 m, a wall-thickness of 19 mm and which is constrained longitudinally. If, in error, a valve in the pipeline is suddenly closed, determine, from first principles, the pressure-rise which will be expected at the valve. What will be the corresponding increase in pipe-circumferential stress?

For steel, $E = 200\ \text{GN/m}^2$.
For water, $K = 2\ \text{GN/m}^2$.

7. A long water-supply line to a service reservoir terminates in an automatically operated valve. It is found that, during slow closure, noisy vibrations ("water hammer") first appear and then die away.

(*a*) Define *slow* and *rapid* closure of the valve, in terms of the length of the pipeline, etc.

(*b*) Derive general formulae for maximum and minimum pressure at the valve, assuming (*i*) slow, (*ii*) rapid, closure, and in both cases a thick, virtually rigid pipeline.

(*c*) Describe the type of defect in the valve which, at a certain setting, will permit self-sustained pressure waves.

8. (*a*) A steel pipe of 1·8 m diameter conducts water at a maximum pressure of 4·8 MN/m^2. The tensile stress in the steel is specified as 93 MN/m^2. Calculate the plate thickness of the (circular) pipe.

(*b*) What non-circular cross-section of pipe might with advantage be adopted for conveying water at a low pressure, and why?

CHAPTER 5

DIMENSIONAL ANALYSIS

DIMENSIONS

VELOCITY—for example—can be measured in m/s, mm/s, or km/h, all of which, although they belong to different systems of units, have in common the property of being a length divided by a time. Because velocity, whatever the units, always has this property it is said to have the *dimensions* of Length/Time, usually written L/T or LT^{-1}. Mass has the dimension of mass, written M. A fourth *prime concept* (as M, L and T are called) is temperature θ but, apart from expressions concerned with heat transfer, all the dimensions of the other quantities encountered in Fluid Mechanics can be expressed in terms of these three, M, L, T, and may be derived from any proper equation containing them. As an example, the dimensions of Force can be derived from Newton's second law:

$$\text{Force} = \text{Mass} \times \text{Acceleration}$$

The dimensions of acceleration (measured in m/s^2, mm/s^2, etc.) are obviously L/T^2.

$$\therefore \text{ the dimensions of force are } ML/T^2$$

The dimensions of other common quantities can be determined in the same way, and are given in the list below.

Mass	M
Length	L
Time	T
Velocity	LT^{-1}
Acceleration	LT^{-2}
Force	MLT^{-2}
Pressure	$ML^{-1}T^{-2}$
Density	ML^{-3}
Specific weight	$ML^{-2}T^{-2}$
Specific gravity	1 (*i.e.* the dimensions of unity)
Energy	ML^2T^{-2}
Momentum	MLT^{-1}
Dynamic viscosity	$ML^{-1}T^{-1}$
Kinematic viscosity	L^2T^{-1}
Hydraulic gradient	1
Hydraulic mean depth	L

A fundamental rule concerning equations is that every term in an equation must have the same dimensions; and it is in the application of this rule that dimensions can be so useful in theory. An example, Chézy's formula, is mentioned on p. 168. Chézy's formula may be rearranged thus:

$$C = u/\sqrt{mi}$$

the dimensions of the right hand side are $L/T \div \sqrt{L} = L^{\frac{1}{2}}T^{-1}$ and consequently C has the dimensions $L^{\frac{1}{2}}/T$.

A knowledge of dimensions, therefore, could safeguard against the mistakes that might otherwise arise when using Chèzy's formula in a problem with unfamiliar units. The term *dimensional analysis*, however, is usually reserved for the technique illustrated in Example 19 in which consideration of the dimensions of the variables involved leads to (what might be described as) a first approximation to the formula connecting them.

EXAMPLE 19

Find by dimensional analysis an expression for the drag force, F, experienced by a sphere of diameter D when moving with a velocity v through a fluid of density ρ and dynamic viscosity μ.

SOLUTION

Using ϕ to denote "a function of" (a common device in mathematics), the given information can be summarised thus:

$$F = \phi(D, v, \rho, \mu) \tag{1}$$

which means "F is a function of D, v, ρ and μ," or (in other words) F depends on D, v, ρ and μ and no other properties. The researchėr tackling this problem for the first time would not know whether the right-hand side contained one term only or more than one. Since dimensions only are to be considered, and since every term (if there are more than one) must have the same dimensions, it is permissible to replace the right-hand side by the general term $kD^a v^b \rho^c \mu^d$, in which k, a, b, c and d are pure numbers. So

$$F = kD^a v^b \rho^c \mu^d \tag{2}$$

The dimensions of both sides must be the same,

i.e. $$MLT^{-2} = (L)^a(LT^{-1})^b(ML^{-3})^c(ML^{-1}T^{-1})^d$$

The power (or *index*) of M must be the same on both sides, as must be the indices of L and T. Equating these indices yields three *indicial* equations, namely

$$\left.\begin{aligned} M&: & 1 &= c + d \\ L&: & 1 &= a + b - 3c - d \\ T&: & -2 &= -b - d \end{aligned}\right\} \tag{3}$$

which can be solved to give expressions for any three of the unknowns a,

b, c, d, in terms of the fourth. Here a, b and c are determined in terms of d:

$$\left.\begin{aligned} c &= 1 - d \\ b &= 2 - d \\ \text{and} \qquad a &= 1 - b + 3c + d = 2 - d \end{aligned}\right\} \tag{4}$$

$$\therefore \qquad F = kD^{2-d}v^{2-d}\rho^{1-d}\mu^{d}$$

$$\text{i.e.} \qquad \underline{\underline{F = kD^2v^2\rho(\rho vD/\mu)^{-d}}} \tag{5}$$

Alternative methods of writing the answer are:

$$\frac{F}{D^2v^2\rho} = k\left(\frac{\rho vD}{\mu}\right)^{-d} \tag{6}$$

$$\text{or} \qquad \frac{F}{D^2v^2\rho} = \phi\left(\frac{\rho vD}{\mu}\right) \tag{7}$$

NOTES

(*i*) Equation (7) is a generalisation of equation (6), again allowing for the possibility of more than one term on the right-hand side. Equations (6) and (7) have the advantage over equation (5) that they present the answer as one group expressed as a function (admittedly unknown) of another group of variables. In practice, dimensional analysis is often carried out as a preliminary to an experiment the object of which is to provide a graph from which an empirical law can be established. Dimensionless groups (such as those in equation (7)) are then the most suitable parameters for plotting.

(*ii*) The procedure—namely collecting all the variables into an equation such as (2), equating dimensions to form the indicial equations (3) and solving these to give an answer in the form of (5), (6) or (7)—is always the same.

(*iii*) The reader should, however, go through the algebra again, solving the indicial equations in terms of the other unknowns to give the following results:

Expressing a, b, d in terms of c gives: $F = kDv\mu(\rho vD/\mu)^c$
" a, c, d " " " b " $F = k\mu^2/\rho(\rho vD/\mu)^b$
" b, c, d " " " a " $F = k\mu^2/\rho(\rho vD/\mu)^b$

At first sight these answers do not seem to agree with equation (5).

(*iv*) All the answers are, in fact, interchangeable. For example, if

$$F = kDv\mu(\rho vD/\mu)^c$$

then the right-hand side can be written

$$kDv\mu . \rho vD/\mu . (\rho vD/\mu)^{c-1}$$

$$\text{so that} \qquad F = kv^2D^2\rho(\rho vD/\mu)^{c-1}$$

which is the same as equation (5), because $c - 1 = -d$, from equation (4).

Similarly, by taking $\rho^2v^2D^2/\mu^2$ outside the brackets, $k\mu^2/\rho.(\rho vD/\mu)^b$ becomes $kD^2v^2\rho(\rho vD/\mu)^{-d}$, again the same as equation (5), and the following rule emerges: If the answer is expressed in terms of non-dimensional groups (such as in equation (6)) it is always permissible to change it by multiplying one group by any power, positive or negative, of the other (or *another*), if there are more than two groups.

(*v*) The number of dimensionless groups which may be expected to result from a particular dimensional analysis can be calculated from Buckingham's Π theorem, which shows that, if there are n variables in a problem and m dimensions involved, there will be $(n-m)$ dimensionless groups in the answer. Thus, in Example 19 there are five variables (namely F, D, v, ρ and μ) and three dimensions, and therefore there are two groups in the answer, namely $F/(D^2v^2\rho)$ and $\rho vD/\mu$. It cannot be emphasised too strongly that these groups are dimensionless, for this property can be used to check the algebra at the end of an analysis—in so far as a mistake must have occurred if one or more groups is found to have dimensions.

(*vi*) One form of the well known Stokes' formula is $F = 6\pi r\mu v$, giving the force F resisting the motion of a small sphere moving slowly through a fluid. It is instructive to compare this formula with $F = kDv\mu(\rho vD/\mu)^c$ obtained above. Since the diameter D is twice the radius r of the sphere, it would appear that $k(\rho vD/\mu)^c = 3\pi$; an experiment would verify this deduction, providing the velocity was low enough. As mentioned on p. 25 $(\rho vD/\mu)$ is called the Reynolds number, R_e, for flow past a sphere, and a series of experiments at different velocities would show that $k(\rho vD/\mu)^c = 3\pi$ only if $R_e < 1$. If $R_e \gg 1$ Stokes' formula no longer determines the drag force accurately, the discrepancy between the value obtained by Stokes' formula and the true drag increasing as R_e increases. In this instance, therefore, dimensional analysis gives a more general result than Stokes' formula, which is obtained by rigorous, advanced mathematics (but which is limited in its scope by the assumptions which have to be made), even though the quantity $k(\rho vD/\mu)^c$ has to be determined experimentally.

(*vii*) Reynolds numbers are defined for a number of flow problems (*see* Example 20), and although the definition varies from problem to problem, a Reynolds number is always of the same form, namely, a density times a velocity times a characteristic length divided by viscosity.

In note (*vi*) the Reynolds number $R_e = 1$ is called the *critical* Reynolds number owing to the fact that flow past a sphere (which is the same hydrodynamically as a sphere moving through a fluid) obeys Stokes' formula or a different formula according to whether Reynolds number is less than or greater than 1. This is typical of a number of problems, such as pipe-flow or flow in open channels, in which Reynolds numbers are defined and, in addition, exhibit critical values separating distinctly different types of flow.

In the moving-sphere problem for a given liquid and a given sphere (so that ρ, μ, and D are fixed), the critical Reynolds number determines a particular velocity; this velocity is called the *critical velocity*. In fact, in all flow problems the velocity corresponding to the critical Reynolds number is always called the critical velocity.

EXAMPLE 20

Define the Reynolds numbers of:

(*a*) a circular pipe;
(*b*) a filter bed consisting of uniform sand grains;

(*c*) a sphere suspended in a stream; and
(*d*) a thin plate inserted longitudinally in a wind tunnel.

Discuss the significance and the utility of this number in connection with one of the four applications listed.

SOLUTION

In the following four parts, ρ and μ are the fluid density and viscosity respectively.

(*a*) For pipe-flow, $R_e = \rho ud/\mu$, where d is the pipe diameter and u is the mean velocity of flow obtained by dividing the volumetric rate of flow through the pipe by its cross-sectional area; *i.e.* $u = Q/A$.

(*b*) For a filter bed, $R_e = \rho ud/\mu$, where d is the mean grain diameter and u is the discharge velocity. Since sand grains are unlikely to be perfectly spherical, the "diameter" of a grain is defined as the diameter of the sphere which, were it to replace the sand grain, would allow the same flow past it. Discharge velocity is defined more fully in *Soil Mechanics*, in this series, and may be calculated by dividing the total discharge from the filter bed by the gross area of the bed.

(*c*) Reynolds number for a sphere suspended in a stream is precisely the same as for the sphere in Example 19, the two problems being identical (since the sphere falls relative to the fluid, and therefore the fluid flows relative to the sphere).

(*d*) For the thin plate, $R_e = \rho ul/\mu$, where l is the length of the plate in the direction of flow and u is the undisturbed air velocity, *i.e.* the velocity of flow well upstream of the plate, *see* Chapter 10. It is necessary to specify the velocity of the undisturbed air because the introduction of even a *thin* plate into the flow will cause local variations in air speed.

The significance of Reynolds number in connection with pipe-flow is discussed in Chapter 11, starting on p. 178.

DIMENSIONLESS NUMBERS EXPRESSED AS FORCE RATIOS

A Reynolds number can be thought of as a measure of the significance of viscosity because, relative to the other quantities involved, a small value of μ gives a large value of R_e, and vice versa. Other dimensionless numbers are in common use, and these are defined and expressed as the ratio of two forces, below. The Froude number, F_r for example, is widely used in connection with channel flow, it being a measure of the influence of gravity on a flow.

In the definitions of these and other dimensionless numbers the characteristic length is given the symbol d. It must be realised that in some problems, however, the symbol l is more convenient, as in Example 20 (*d*), and that in general any length can be used. In practice, the choice of the characteristic length is a matter of convention and convenience.

Reynolds number, $R_e = \rho ud/\mu$ = Inertia force/Viscous force.

Froude number, $F_r = u^2/(gd)$ = Inertia force/Gravitational force. A Froude number is often defined as the square root of this expression, namely $u/\sqrt{gd}$, but both forms are used and the given form is the more convenient here. The use of the second form is illustrated on p. 79.

Weber number, $W_e = u^2\rho d/\sigma$ = Inertia force/Surface tension force, where σ = surface tension, expressed in force/unit length. This number, too, is sometimes defined by the square root, namely $u\sqrt{\rho d/\sigma}$, and elsewhere by the reciprocals of these groups. The definition $u^2\rho d/\sigma$ is chosen here as being the most consistent with the definitions of the other dimensionless numbers. Although the variety of definitions is apt to be bewildering at first, the student will find that, in most serious work on fluid mechanics, any dimensionless parameter(s) used is explicitly defined.

Mach number, $M_a = u/\sqrt{K/\rho}$ = Inertia force/Elastic force, where K = bulk modulus of elasticity of the fluid. The Mach number is a measure of the effects of compressibility and becomes an important consideration only at high velocities, *e.g.* when aircraft speeds are approaching the velocity of sound, as is well known. In fact, $\sqrt{K/\rho}$ is the velocity of sound in a fluid with the properties K and ρ, so that an alternative way of defining M_a is:

$$M_a = \text{Flow velocity/Velocity of sound.}$$

The definitions of these dimensionless numbers as force ratios may be proved from the following:

(*a*) Inertial reaction = Mass × Acceleration.

and Acceleration $\propto u^2/d$, which has the same dimensions

$$\text{Mass} = \rho \times \text{Volume} \propto \rho d^3$$

$$\therefore \text{Inertia force} \propto \rho d^2 \times u^2/d = \rho d^2 u^2$$

(*b*) Viscous force = Viscous stress × Area

$$\propto \mu\frac{du}{dy} \times d^2 \propto \mu u/d \times d^2 = \mu u/d$$

(*c*) Gravitational force = Mass × g $\propto \rho d^3 g$.

(*d*) Surface tension force = σ × Length $\propto \sigma d$.

(*e*) Elastic force = K × area $\propto Kd^2$, since K = stress/strain and therefore has the dimensions of stress.

Dividing (*a*) by (*b*) gives Reynolds number, which may therefore be thought of as an inertia force divided by a viscous force. Similarly, dividing (*a*) by (*c*) gives a Froude number; dividing (*a*) by (*d*) gives a Weber number; and dividing (*a*) by (*e*) gives a Mach number.

The conception of these dimensionless numbers as force ratios provides the basis of dynamical similarity, which is dealt with in Chapter 6.

Example 21 illustrates how both the Reynolds number and the Froude number (this time in the form $u/\sqrt{dg}$) can occur in the same analysis. The technique used is exactly the same as in Example 19, except that it is extended slightly to yield a particular result.

EXAMPLE 21

If it is believed that the volumetric discharge Q through an orifice may be dependent on the fluid coefficient of dynamic viscosity μ, the fluid density ρ, the orifice-diameter d, the head over the orifice H and gravitational acceleration g show by means of a dimensional analysis that:

$$Q = C_D A\sqrt{2gh}$$

where A is the area of the orifice and C_D is a function of the two ratios: Froude number/Reynolds number and H/d.

Comment briefly on the importance and limitations of dimensional analysis.

SOLUTION

Summarising the information given in the question,

$$Q = \phi(\mu, \rho, d, H, g)$$

$\therefore$ let

$$Q = k\mu^a \rho^b d^c H^e g^f$$

The dimensions must be the same on both sides;

i.e.

$$L^3T^{-1} = (ML^{-1}T^{-1})^a(ML^{-3})^b(L)^c(L)^e(LT^{-2})^f$$

and the indicial equations are:

$$M:\quad 0 = a + b \qquad (1)$$

$$L:\quad 3 = -a - 3b + c + e + f \qquad (2)$$

$$T:\quad -1 = -a - 2f \qquad (3)$$

$\therefore$ from (1) $\quad b = -a$

from (3) $\quad f = \frac{1}{2}(1 - a)$

from (2) $\quad 3 = -a + 3a + c + e + \frac{1}{2}(1 - a)$

$\therefore \quad c = 5/2 - 3a/2 - e$

Hence,

$$Q = k\mu^a \rho^{-a} d^{(\frac{5}{2} - \frac{3}{2}a - e)} H^e g^{\frac{1}{2}(1-a)}$$

$$= kd^{\frac{5}{2}}g^{\frac{1}{2}}\left(\frac{\mu}{\rho d^{\frac{3}{2}} g^{\frac{1}{2}}}\right)^a (H/d)^e$$

Now $$\mu/(\rho d^{\frac{3}{2}}g^{\frac{1}{2}}) = \mu/(\rho v d) \times u/(d^{\frac{1}{2}}g^{\frac{1}{2}}) = F_r/R_e$$

and $$d^{\frac{5}{2}}g^{\frac{1}{2}} = d^2\sqrt{Hg} \times (d/H)^{\frac{1}{2}}$$

$\therefore$ $$Q = kd^2\sqrt{gH}(F_r/R_e)^a(H/d)^{(e-\frac{1}{2})}$$

Since A = constant $\times d^2$,

$$Q = A\sqrt{2gH}\,.\,\phi(F_r/R_e, H/d) \tag{4}$$

where the function ϕ includes all the constants. Also, if

$$C_D = \phi(F_r/R_e, H/d)$$

then $$Q = C_D A\sqrt{2gH}$$

The importance and limitations of dimensional analysis are discussed in notes (*i*) and (*vi*) to Example 19.

NOTES

(*i*) The change from $kd^2(F_r/R_e)^a(H/d)^{(e-\frac{1}{2})}$ to $2A\phi(F_r/R_e, H/d)$ in this example is convenient, as it saves the extra arithmetic involved by introducing the constants $\pi/4$ and $\sqrt{2}$, which would otherwise be necessary. The solution could, however, be completed using the first form of expression (characterised by the constants k, a, $e-\frac{1}{2}$, etc.) throughout.

(*ii*) In this example six variables and three dimensions are involved and therefore, by Buckingham's Π theorem, there are three dimensionless groups in the answer, namely $Q/(d^2\sqrt{gH})$, $\mu/(\rho d^{\frac{3}{2}}g^{\frac{1}{2}})$ or F_r/R_e, and H/d. This is a useful check which should always be made.

Moreover, in connection with this rule, it should be noticed that the three indicial equations relate five unknown indices, which makes it possible to solve for any three in terms of the other two.

QUESTIONS

1. If a submerged weir is built across a small river, show, by dimensional analysis, that the force F on it due to the flow of the water is

$$F = \rho u^2 d^2\,\phi(ud^{-\frac{1}{2}}g^{-\frac{1}{2}}$$

where ρ = density of the water;

u = mean velocity of flow; and

d = depth of the weir.

2. If, in question 1, the force F on the weir depends on the coefficient of viscosity, μ, in addition to ρ, u, d and g, show that

(*i*) $F = \rho u^2 d^2\,\phi(\rho u d/\mu,\ u/\sqrt{gd})$

(*ii*) $F = k\rho u^2 d^2[\mu/(\rho u d)]^e[u/\sqrt{gd}]^{2f}$

3. Reynolds and Froude numbers each represent ratios of forces influencing fluid motions. Define these force ratios and hence, in each case, determine the conventional expression for the ratio.

Explain why these dimensionless numbers, or force ratios, are so important in engineering practice and illustrate clearly, with examples, their distinctive applications.

4. If the resistance to flow through a pipeline is expressed as τ (force per unit area of pipe-wall) and is assumed to depend on the mass density of the fluid ρ, the mean velocity of flow u, the coefficient of dynamic viscosity μ, the pipe diameter d, and the height of the wall-roughness k, show (by means of an analysis of the dimensions) that $\tau/(\rho u^2)$ can be expressed as a function of the two quantities (k/d) and $(\rho ud/\mu)$.

Comment on the significance of this analysis and show that the result may be related to the Darcy equation

$$h_f = 4f\,lu^2/2gd$$

[*Hint*. In the first part, prove $\tau/(\rho u^2) = \phi(\rho ud/\mu, k/d)$. In the second part the f of Darcy's equation turns out to be a function of Reynolds number and k/d.]

5. Define two of the following parameters: Reynolds number, Froude number, Mach number.

In both cases quote "critical" numerical values and briefly describe the change occurring in fluid flow at or near those values.

6. The wave force exerted on a small diameter pile in the sea is given by

$$F = C_m \rho B a$$

Show that C_m, the dimensionless coefficient of mass, is a function of $\dfrac{d\omega^2}{a}$ and $\dfrac{L}{d}$, if it may be assumed that $\phi[F, \omega, a, \rho, L, d] = 0$, F is the instantaneous wave force, and:

$\omega = 2\pi/T$ where T is the wave period;
a = wave particle acceleration;
ρ = the fluid density;
B = the submerged volume of the pile;
L = the submerged length of the pile; and
d = the pile diameter.

CHAPTER 6

DYNAMICAL SIMILARITY

MANY engineering systems, such as hydro-electric schemes, process plant and fluid machines, involve types of flow which, because of their complexity, cannot be analysed satisfactorily by mathematics. When a contract involves a great amount of money it is usually economical to construct a scale model and perform tests on it in order to predict what will happen in the prototype (*i.e.* the full-sized system). The interpretation of model test results, however, is a highly skilled business and calls for experience as well as a complete understanding of the principles of dynamical similarity. The idea of dynamical similarity is best thought of as a development of dimensional analysis and, as it is not easy to paraphrase the statement that two flow systems are dynamically similar, the principles are established gradually throughout this chapter.

CONCEPT OF A MODEL

Supposing that a new process device has been designed. The device incorporates "forced flow", *i.e.* the fluid is driven through the system by a machine such as a fan or pump. The Power P required to drive the machine is assumed to be related to the size D and rotational speed n of the machine, and also to the fluid properties (velocity u, density ρ and kinematic viscosity ν). Since the system must achieve the highest possible efficiency it has been decided that model tests should be performed. Answers must, therefore, be found to some or all of the following questions:

(*a*) What is the correct velocity of flow for the model?
(*b*) Consequently what is the correct volumetric flow rate in the model?
(*c*) How will the power measured on the model be related to the power for the full-scale prototype?
(*d*) What is the correct rotational speed for the model of the machine?
(*e*) If the process involves fluctuations with time, how does the time scale for the model compare with time scale for the prototype?

Dimensional analysis provides the answers.

Assuming that the power depends on the other system parameters, then:

$$P = \phi(\rho, v, D, u, n) \tag{1}$$

$\therefore$ let

$$P = k\rho^a v^b D^c u^d n^e$$

The dimensions must be the same on both sides;

i.e. $$ML^2T^{-3} = (ML^{-3})^a(L^2T^{-1})^b(L)^c(LT^{-1})^d(T^{-1})^e$$

and the indicial equations are:

$$\begin{aligned} M: &\quad 1 = a \\ T: &\quad -3 = -b - d - e \\ L: &\quad 2 = -3a + 2b + c + d \end{aligned}$$

$\therefore$ $$a = 1,\ b = 3 - d - e,\ c = d + 2e - 1$$

$\therefore$ $$P = k\rho v^3 D^{-1}(Duv^{-1})^d(nD^2v^{-1})^e \tag{2}$$

This analysis applies equally well to prototype and model. Hence, using the suffix 1 to denote properties of the prototype and the suffix 2 to denote properties of the model:

$$\frac{P_1}{\rho_1 v_2{}^3 D_1{}^{-1}} = k(D_1 u_1 v_1{}^{-1})^d(n_1 D_1{}^2 v_1{}^{-1})^e \tag{3}$$

$$\frac{P_2}{\rho_2 v_2{}^3 D_2{}^{-1}} = k(D_2 u_2 v_2{}^{-1})^d(n_2 D_2{}^2 v_2{}^{-1})^e \tag{4}$$

The unknown constant k can be eliminated by dividing equation (2) by equation (3), but the resulting equation is of no use unless, in addition,

$$(D_1 u_1 v_1{}^{-1})^d = (D_2 u_2 v_2{}^{-1})^d \tag{5}$$

and

$$(n_1 D_1{}^2 v_1{}^{-1})^e = (n_2 D_2{}^2 v_2{}^{-1})^e \tag{6}$$

so that the unknown indices, d and e may also be eliminated. Then

$$\frac{P_1}{\rho_1 v_1{}^3 D_1{}^{-1}} = \frac{P_2}{\rho_2 v_2{}^3 D_2{}^{-1}} \tag{7}$$

The groups (Duv^{-1}) and (nD^2v^{-1}) are both forms of the Reynolds number R_e. The first group is the R_e describing the fluid flow and the second is the R_e describing the action of the machine. It has already been stated that Reynolds number represents a ratio of forces; therefore, for equations (5) and (6) to be true it is necessary that the ratio of forces in the prototype be equal to the ratio of forces in the model. This is the essence of "dynamical similarity". A true model will also be geometrically similar to the prototype, so that the ratio of any corresponding lengths in prototype and model will be equal to D_1/D_2. The relationship between the cross-sectional area in the prototype and the corresponding area in the model is therefore

$$a_1/a_2 = (D_1/D_2)^2$$

If the fluid properties ρ, ν are known for the prototype and for the model, equation (5) may be used to calculate the velocity scale factor for the flow:

$$\frac{u_1}{u_2} = \frac{D_2 \nu_2^{-1}}{D_1 \nu_1^{-1}}$$

u_1 and u_2 are known as "corresponding velocities"; similarly equation (6) may be used to calculate the scale factor for the corresponding rotational speeds:

$$\frac{n_1}{n_2} = \frac{D_2 \nu_2^{-1}}{D_1 \nu_1^{-1}}$$

Since distance divided by velocity has the dimensions of time, a time scale may be found:

$$\frac{t_1}{t_2} = \frac{D_1 u_1^{-1}}{D_2 u_2^{-1}}$$

The foregoing text should be studied carefully, for it shows the procedure which can always be followed in problems on dynamical similarity, namely:

(*a*) the formation of the initial equation (equation (1))

(*b*) a dimensional analysis leading to two equations (such as (3) and (4)) one for the prototype and one for the model.

(*c*) the equating of the groups of variables which are raised to unknown powers.

(*d*) the equating of the remaining groups (equation (6)).

Effect of including further parameters on the dimensional analysis

In the above analysis it was assumed that the power was a function of ρ, u, D and ν only. If the list is extended to include say, surface tension σ and gravitational acceleration g, then equation (1) becomes:

$$P = \phi(\rho, \nu, D, u, n, \sigma, g)$$

and dimensional analysis gives

$$P = \rho \nu^3 D^{-1} \phi(R_e, F_r, W_e) \qquad (8)$$

the proof is left to the reader.

Equation (8) raises a difficulty which is frequently encountered in modelling. In order to achieve a true dynamical similarity it is necessary to satisfy all three of the dimensionless numbers (R_e, F_r, W_e) simultaneously for both the model and the prototype. That is to say the model and prototype are only behaving in a dynamically similar fashion if the ratio P_1/P_2 is the same at all geometrically corresponding points in the model and the prototype. This is true only if:

R_e has the same value at corresponding points in the model and the prototype;
F_r has the same value at corresponding points in the model and the prototype;
W_e has the same value at corresponding points in the model and the prototype.

If the flow were compressible the above list would have to be extended to include M_a.

All these dimensionless groups represent force ratios and it can be reasoned that similar conclusions can be drawn for all flow problems. Now the difficulty is that it is almost always impossible to satisfy simultaneously the requirements of all the dimensionless groups. However, in most practical cases it either happens or is contrived that one of the dimensionless groups is far more important than the others. This idea will now be illustrated.

EXAMPLE 22

A submerged weir is to be built across a small river of unusual cross section. The influence of the weir is to be determined before its construction, from studies made with a hydraulic model which has a linear scale of 1:20. Determine the corresponding scales of velocity, discharge, time, pressure and force for this model.

SOLUTION

It is assumed that the force of the river depends on ρ, u, d and g, and therefore

$$F = k\rho^a u^b d^c g^e$$

Dimensional analysis leads to the result

$$F = \rho u^2 d^2 k(ud^{-1/2} g^{-1/2})^a$$

Hence if the suffix 1 relates to the prototype and the suffix 2 to the model

$$F_1 = \rho_1 {u_1}^2 {d_1}^2 k(u_1 {d_1}^{-1/2} g^{-1/2})^a$$
$$F_2 = \rho_2 {u_2}^2 {d_2}^2 k(u_2 {d_2}^{-1/2} g^{-1/2})^a$$

We say, for dynamical similarity

$$(u_1 {d_1}^{-1/2} g^{-1/2}) = (u_2 {d_2}^{-1/2} g^{-1/2})$$

$\therefore$ since $g_1 = g_2$, $u_1/u_2 = (d_1/d_2)^{1/2} = \sqrt{20} = \underline{\underline{4.48}}$

As water will be used for the model, $\rho_1 = \rho_2$

$$\therefore \quad F_1/F_2 = u_1{}^2 d_1{}^2/u_2{}^2 d_2{}^2 = (d_1/d_2)^3 = 20^3 = \underline{\underline{8000}}$$

Since $Q = ua$, the discharge scale factor is

$$\therefore \quad Q_1/Q_2 = (u_1/u_2)(d_1/d_2)^2 = (d_1/d_2)^{5/2} = 20^{5/2} = \underline{\underline{1790}}$$

The time scale factor is given by

$$t_1/t_2 = (d_1/u_1)/(d_2/u_2) = (d_1/d_2)^{1/2} = \sqrt{20} = \underline{\underline{4.48}}$$

Since pressure = force/area, scale factor is therefore

$$P_1/P_2 = (F_1/d_1{}^2)/(F_2/d_2{}^2) = d_1/d_2 = \underline{\underline{20}}$$

It was assumed above that the force depended only on ρ, d, u and g. Obviously under some conditions viscous drag will not be negligible so let us see the effect on the dimensional analysis of introducing μ, the coefficient of dynamic viscosity.

This time the initial equation is

$$F = \phi(\rho, u, d, \mu, g)$$

Dimensional analysis yields the result

$$F = \rho u^2 d^2 (\mu/(\rho u d))^a (u/(gd)^{1/2})^{2b} \tag{9}$$

Therefore for dynamic similarity two conditions must be satisfied, namely:

$$\mu_1/(\rho_1 u_1 d_1) = \mu_2(\rho_2 u_2 d_2) \tag{10}$$

and
$$u_1/(gd_1)^{1/2} = u_2/(gd_2)^{1/2} \tag{11}$$

If the same fluid (water) is used in the model as in the prototype the $\rho_1 = \rho_2$ and $\mu_1 = \mu_2$, $\therefore$ equation (10) reduces to

$$u_1 d_1 = u_2 d_2 \tag{12}$$

and as g is common to both sides equation (3) reduces to

$$u_1/d_1{}^{1/2} = u_2/d_2{}^{1/2} \tag{13}$$

The only solution which would simultaneously satisfy both equations is $d_1 = d_2$ and would require the model to be of the same size as the prototype! It is evident, therefore, that true dynamical similarity cannot be obtained for flows influenced by two or more force systems (viscous and gravity forces in this example). Model analysis is, however, still used in these circumstances by satisfying equation (13) and ignoring (12). This approach works well in practice, since most flow problems are influenced predominantly by one force system, flows over weirs or down spillways being primarily influenced by gravity. It should also be noted that equation (13) gives a convenient velocity scale, namely $u_1/u_2 = (d_1/d_2)^{1/2}$, whereas equation (12) leads to the scale $u_1/u_2 = d_2/d_1$, which makes the model velocity extremely high. The following problem illustrates an approach sometimes used if equation (12) is the dominant one.

EXAMPLE 23

Fluid motion may be influenced by forces from differing origins. Outline these various forces (and their origins) and state the conditions which ideally must be imposed in order to reproduce a particular fluid motion at a reduced scale.

Water, with a coefficient of kinematic viscosity of 1.12×10^{-6} m²/s and a mass density of 1 Mg/m³, flows at a mean speed of 1.75 m/s through a 75 mm diameter pipeline. What corresponding volumetric rate (measured at atmospheric pressure) of air-flow through this pipeline would give rise to essentially dynamically similar flow conditions and why would this be so? Air under the test conditions may be assumed to have a coefficient of kinematic viscosity of 14.7×10^{-6} m²/s and a mass density of 1.23 kg/m³.

Determine, for each fluid, the pressure drop (N/m²) which would occur in a 10 m length of this pipeline. Assume, for the conditions specified, that $f = 0.010$ for the pipeline and that the influence of changes in the air density in the pipeline are negligible.

SOLUTION

The forces and their origins are given on pp. 71 and 72, and the conditions are given on p. 79; they are that F_r, R_e, W_e, M_a should each have one value for both model and prototype.

In the numerical example the flow will be influenced mainly by viscosity, and therefore the criterion for dynamical similarity is that Reynolds number should be the same for both the water-flow and the air-flow. (In questions of this nature such reasoning is quite permissible; the alternative approach is to perform a dimensional analysis which leads to the same observation.)

Kinematic viscosity, $v = \mu/\rho$, and therefore, $R_e = ud/v$

$$\text{for water, } R_e = 1{\cdot}75 \times 75 \times 10^{-3}/(1{\cdot}12 \times 10^{-6})$$

$$= 1{\cdot}172 \times 10^5$$

$$\therefore \quad \text{for air, } R_e = 1{\cdot}172 \times 10^5$$

$$= u \times 75 \times 10^{-3}/(14{\cdot}7 \times 10^{-6})$$

$$\therefore \quad \text{the air velocity, } u = 23 \text{ m/s}$$

The volumetric rate of air flow $= 23 \times \pi/4 \times 75^2 \times 10^{-6}$

$$= 0{\cdot}1017 \text{ m}^3/\text{s}$$

By Darcy's formula (see p. 30), the pressure drop in a length l of pipe is given by

$$p = \frac{4flu^2}{2gd}\rho g \text{ N/m}^2$$

if all the quantities are measured in SI units.

Hence, for the water-flow

$$p = \frac{4 \times 0{\cdot}010 \times 10 \times 1{\cdot}75^2 \times 10^3}{2 \times 75 \times 10^{-3}} = 8{\cdot}17 \text{ kN/m}^2$$

and for the air-flow

$$p = \frac{4 \times 0{\cdot}010 \times 10 \times 23^2 \times 1{\cdot}23}{2 \times 75 \times 10^{-3}} = 1{\cdot}73 \text{ kN/m}^2$$

APPLICATION TO FLUID MACHINES

The concepts of dimensional analysis and dynamical similarity may be applied to fluid machines such as pumps or fans. For example the functional relationship between the various dimensionless ratios for pumps is usually expressed in the form:

$$\frac{P}{\rho N^3 D^5} = \phi\left(\frac{Q}{ND^3}, \frac{gH}{N^2 D^2}, \frac{\rho N D^2}{\mu}, \frac{k}{D}\right) \tag{14}$$

P = power supplied (or extracted), N = rotational speed, k = roughness, and $\therefore$ k/D is a relative roughness. The other letters have their usual meaning.

$\dfrac{P}{\rho N^3 D^5}$ is known as the *power constant*, K_P

$\dfrac{Q}{ND^3}$ is the *capacity constant*, K_Q

$\dfrac{gH}{N^2 D^2}$ is the *head constant*, K_H

$\dfrac{\rho N D^2}{\mu}$ is obviously a form of the Reynolds number.

These dimensionless groups or "constants" may be applied not only for model tests but also:

(*a*) to use data obtained from an existing machine to predict the performance of a new design. The new machine would of course, have to be of similar form to the existing one.

(*b*) to predict the performance of the existing machine under different operating conditions such as a change of speed (N).

The pump industry does not always use machine constants in exactly the above form. The "specific speed" K_s for example is derived by taking the ratio $(K_Q)^{1/2}/(K_H)^{3/4}$, which gives:

$$K_s = N\frac{Q^{1/2}}{(gH)^{3/4}}$$

This is the correct dimensionless form, but because gravitational acceleration, g, is constant it is often omitted. To distinguish this second form of specific speed it will be denoted N_s; naturally this form is not dimensionless.

The value of K_s or N_s for a particular machine is calculated for its designed optimum operating duty. The term "specific speed" is unfortunate, as neither K_s nor N_s represent a machine speed. The point of the specific speed is that its numerical value is a constant for a "family" of similar machines. It is, therefore, best to think of it as a factor which is related to the geometrical form of a machine, and to its designed duty.

Although this analysis and discussion has centred on pumps, the same principles apply for all fluid machines. However, for a number of reasons the dimensionless machine constants may appear in various forms. For example, the specific speed for pumps is based on *its* primary characteristic, Q, whereas for water turbines the principal performance characteristic of the machine is its power. The specific speed of a turbine therefore relates power speed and head thus:

$$K_s = \frac{NP^{1/2}}{\rho^{1/2}(gH)^{5/4}} \qquad \text{or} \qquad N_s = \frac{NP^{1/2}}{H^{5/4}}$$

For those machines in which a compressible fluid is used it is usually appropriate to use the mass flow rate $\dot{m}$ rather than volumetric flow Q.

EXAMPLE 24

A centrifugal pump has an impeller of 500 mm diameter which rotates at 1450 rev/min. Its designed duty is as follows:

Discharge, 120 l/s water ($\rho = 1000$)

Head, 15 m

Power consumption = 20 kW

Estimate the performance of a similar pump with an impeller of 750 mm dia. and with the same rotational speed. Calculate also the specific speed N_s for both pumps.

Subscripts 1 and 2 represent the 500 mm and 750 mm diameter pumps, respectively.

SOLUTION

$$K_{Q1} = K_{Q2} \therefore \frac{0{\cdot}12}{1450 \times 0{\cdot}5^3} = \frac{Q_2}{1450 \times 0{\cdot}75^3} \therefore Q_2 = 0{\cdot}405 \text{ m}^3/\text{s}$$
$$= \underline{\underline{405 \text{ l/s}}}$$

$$K_{H1}=K_{H2} \quad \therefore \quad \frac{9{\cdot}81\times 15}{1450^2\times 0{\cdot}5^2}=\frac{9{\cdot}81\times H_2}{1450^2\times 0{\cdot}75^2} \quad \therefore \quad H_2=\underline{\underline{33{\cdot}75}}$$

$$K_{P1}=K_{P2} \therefore \frac{20000}{1000\times 1450^3\times 0{\cdot}5^5}\times\frac{P_2}{1000\times 1450^3\times 0{\cdot}75^5} \quad P_2=151875$$

say $\underline{\underline{152 \text{ kW}}}$

$$N_{s1}=\frac{1450\times 0{\cdot}12^{1/2}}{15^{3/4}}=\underline{\underline{65{\cdot}9}}$$

$$N_{s2}=\frac{1450\times 0{\cdot}405^{1/2}}{33{\cdot}75^{3/4}}=\underline{\underline{65{\cdot}9}}$$

The above example illustrates how to calculate the design performance of a larger machine using corresponding data from the smaller machine. If a full set of performance curves were available for the smaller machine it would be possible to produce the curves for the larger machine, by simply reading more values of P, Q, and H, from the performance curves and then repeating the above calculations. Typical performance curves for pumps are illustrated on p. 204.

QUESTIONS

1. The resistance to motion of a hydroplane may be considered as due to wave making (or gravity) resistance only. Derive the expression for this resistance,

$$R=\rho u^2 l^2\,\phi(u\sqrt{lh})$$

where R is the resistance and the other symbols have their usual meanings.

A hydroplane is to run at 60 km/h in sea water. Determine the speed at which a 1:5 scale model should be run in fresh water to determine the resistance of the hydroplane, and the ratio between the resistance to motion of the model and full-size boat at these speeds.

Take the densities of sea water and fresh water to be 1·026 and 1·000 Mg/m³, respectively.

2. What is a Froude number, and in what circumstances can it be used as a basis for hydraulic model testing? Mention briefly two examples in which tests could usefully be carried out using this criterion.

Given that one of the tests in the examples quoted is to be made on a model with a linear scale of $\frac{1}{30}$ of full size, determine, from first principles, the corresponding scales of velocity, discharge, force and time. Show also that the Froude numbers would be the same.

3. The nets of a trawler are suspended from a submerged boom, of length $1 = 12$ m and diameter $d_1 = 230$ mm, the drag coefficient of which is $C_D = 1{\cdot}2$. In order to reduce the fuel consumption the skipper replaces the 230 mm

boom by an equally long "stream-lined" one of thickness $d_2 = 190$ mm and $C_D = 0{\cdot}4$.

(*a*) Express the drag force D on the old boom in terms of ρ, C_D, v, d and l (the mass density of sea water is $\rho = 1{\cdot}026$ Mg/m^3). State by what method your formula was first derived.

(*b*) Calculate the saving in power at a cruising speed of $v = 2$ m/s resulting from the replacement of the boom.

4. The torque, T, required to rotate a disc which is totally immersed in a fluid is thought to depend on the following parameters:

the disc diameter, D
the rotational speed of the disc, N
the fluid density, ρ
the dynamic viscosity, μ

Using dimensional analysis show that a functional relationship for the torque in terms of the other parameters is

$$T = \rho N^2 D^5 \phi\left(\frac{\mu}{D^2 N \rho}\right)$$

Model tests using a 200 mm diameter disc gave the following results:

Disc diameter = 200 mm
Measured torque = $2{\cdot}4 \times 10^{-3}$ N-m
Rotational speed = 20 rev/s
Fluid: water
Fluid density = 1000 kg/m^3
Dynamic viscosity = $1{\cdot}2 \times 10^{-3}$ kg/ms

The prototype disc is 2 m diameter and will run in oil of density 860 kg/m^3 and viscosity $1{\cdot}0 \times 10^{-2}$ kg/ms. What will be the corresponding speed and torque for the prototype?

5. The following data were obtained from an axial flow pump handling oil of density 800 kg/m^3 and viscosity 2×10^{-3} kg/ms.

Pump diameter = 800 mm, Pump speed 250 rev/min, Pump power 60 kW, Discharge = $1{\cdot}0$ m^3/s, Head rise 7·65 m.

Using similarity techniques predict the corresponding performance data for a 2 m diameter fan. For air $\rho = 1{\cdot}28$ kg/m^3, $\mu = 1{\cdot}8 \times 10^{-5}$ kg/ms.

6. Describe a hydraulic scale model of which you have knowledge: quote the object of the model investigation, the features of the prototype flow which the model would be expected to reproduce, the similarity criterion and the linear and discharge scales.

7. Stokes showed that, at Reynolds numbers, R_e, small enough for inertia forces to be negligible in comparison with viscous forces, the drag D on a sphere of diameter d moving at a terminal velocity U in a fluid of viscosity μ is $D = 3\pi\mu dU$.

(*a*) Show that the drag coefficient for this kind of sphere may be expressed as $C_D = 24/R_e$.

(*b*) Assuming that fine cubical sand grains (density ρ, side length a) in water obey Stokes' law, calculate the model size of sand grains of the same density for a 1·25 undistorted scale model. Assume that the similarity criterion of the model determining the time scale is the Froude number F_r. Explain why in hydraulic models involving sedimentation it is often convenient to use model sediment of lower density than that of the prototype.

CHAPTER 7

PIPE-FLOW

IN problems on pipe-flow it may be assumed that the pipe is running full unless there is a specific statement to the contrary. Pipes which are not running full can be analysed by the methods used for flow in open channels, and therefore do not belong to this chapter. Since the pipes in the problems of this chapter are always full of water, two fundamental observations can be made:

1. In pipes of uniform cross-section (the common case) the mean velocity of the liquid is the same throughout the length of the pipe by virtue of the continuity equation.
2. Flow in a pipe can occur only when the pressure at one end is greater than the pressure at the other end. Moreover, providing the column of water in the pipe is not accelerating (*i.e.* the flow is *steady*), the available head producing flow must equal the sum of the head losses suffered by the flow, by virtue of the energy principle.

It should be remembered that a head of fluid, h, can always be converted into a pressure, ρgh (ρ being the density of the fluid), and vice versa.

LOSSES OF ENERGY IN PIPE-FLOW

Shock losses of energy occur at any discontinuity of an otherwise straight, uniform pipe, for example, at valves, elbows, bends and changes of section, the energy being dissipated by eddies in the form of heat. Figure 40 (*a*) shows a sudden enlargement in a pipeline, and Fig. 40 (*b*) a sudden contraction. The loss at the sudden enlargement may be derived by assuming that the pressure on the shoulder *AB* is the same as the upstream pressure, p_1, and then applying the momentum principle and Bernoulli's equation to the flow between sections *I* and *II*. Notice that when the enlargement is very great (*e.g.* a pipe exit to a reservoir) $u_2 = 0$ and the energy loss becomes $h_L = u_1{}^2/(2g)$. If, in either of these two cases, the enlargement is made gradually by introducing a taper the loss is reduced.

In a sudden contraction little loss occurs between sections *I* and *II* (Fig. 40 (*b*)), where the flow is convergent. At section *II* a *vena contracta* forms of cross-sectional area $a_2 < a_3$ (a_3 = area of the pipe at section *III*) in other words, the stream at section *II* is surrounded by "dead" water, and it is the divergent flow beyond this section in which the energy loss, h_L, is produced. In fact, the expression $h_L = u_3^2(1/C_c - 1)^2/(2g)$ is derived from the result for a sudden enlargement applied between sections *II* and *III*, taking $C_c = a_2/a_3 = u_3/u_2$. At the entrance to a pipe from the reservoir the energy loss is usually taken as $\frac{1}{2}u^2/(2g)$. In either case, however, if the entrance to the smaller pipe is well rounded the loss is reduced virtually to zero, for no *vena contracta* forms. (These and other shock losses are tabulated in most comprehensive textbooks on hydraulics.)

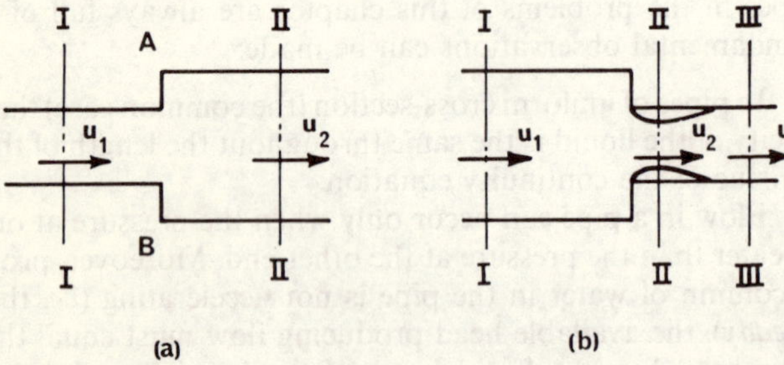

FIG. 40. Pipe-flow: (*a*) sudden enlargement, $h_L = (u_1 - u_2)^2/(2g)$; (*b*) sudden contraction, $h_L = u_3^2\,(1/C_c - 1)^2/(2g)$

In long pipes the shock losses are small compared with the loss of energy due to friction, and can often be neglected altogether, as in Example 26. Friction losses in pipes are described in Chapter 2; in engineering problems the flow is nearly always turbulent and the loss is given by the Darcy formula $h_f = 4flu^2/(2gd)$, *see* p. 30.

Example 25 concerns a short pipe in which the shock losses are not negligible compared with the friction loss. In examples such as this, in which a tank discharges to the atmosphere through a pipe, some of the potential energy available at the tank is converted into kinetic energy, $u^2/(2g)$. Although this is not strictly a loss of energy, the term $u^2/(2g)$ has to be included when equating the total head available with the energy losses.

EXAMPLE 25

The outlet from a water tank consists of a sharp-edged entry to a horizontal pipe of 150 mm diameter, 30 m long, which discharges to the atmosphere through a valve. When fully open the valve causes a head loss of $\frac{1}{4}u^2/(2g)$. The pipe has a (Darcy) friction coefficient of $f = 0{\cdot}006$. What are the limiting improvements in discharge, expressed as a percentage, obtainable: (*a*) by the replacement of the abrupt by a carefully rounded entry, and (*b*) by the removal of the valve?

SOLUTION

The losses of energy in m of water, are:

Friction $= 4fl/d.(\frac{1}{2}u^2/g) = 4 \times 0{\cdot}006 \times 30/0{\cdot}150.(\frac{1}{2}u^2/g) = 4{\cdot}8.(\frac{1}{2}u^2/g)$

Kinetic energy $= 1{\cdot}0(\frac{1}{2}u^2/g)$

Valve $= 0{\cdot}25(\frac{1}{2}u^2/g)$

Abrupt entry $= 0{\cdot}5(\frac{1}{2}u^2/g)$

Total $= 6{\cdot}55(\frac{1}{2}u^2/g)$

hence Total head available $= 6{\cdot}55(\frac{1}{2}u^2/g)$

(*a*) The value often taken for the loss at a well-rounded entry is $0{\cdot}05 \times (\frac{1}{2}u^2/g)$, reducing the total loss to $6{\cdot}10 \times (\frac{1}{2}u_a{}^2/g)$. Hence, if the velocity after this modification is u_a,

$$6{\cdot}55 \times (\tfrac{1}{2}u^2/g) = 6{\cdot}10 \times (\tfrac{1}{2}u_a{}^2/g)$$

from which $u_a/u = 1{\cdot}036$

$\therefore$ the improvement in the velocity (and $\therefore$ in the discharge) is

$$\underline{\underline{3{\cdot}6\%}}$$

(*b*) If, after removing the valve, the velocity $= u_b$, then

$$6{\cdot}30 \times (\tfrac{1}{2}u_b{}^2/g) = 6{\cdot}55 \times (\tfrac{1}{2}u^2/g)$$

$$u_b/u = 1{\cdot}02$$

$$\therefore \text{ improvement} = \underline{\underline{2\%}}$$

EXAMPLE 26

A 130 mm internal-diameter horizontal pipeline, 360 m long, connects two vertical-walled, open, water reservoirs. One of these reservoirs has a plan area of 78 m^2 and, at a given instant, the water-level in it is 6·5 m above the pipe centre-line. At the same instant the water-level in the other, which has a plan area of 50 m^2, is 2·5 m above the pipe centre-line. Find how long it will take for the level in the larger reservoir to fall 1·0 m. Assume $f = 0{\cdot}007$ for the pipeline and disregard all losses other than those in the pipeline.

SOLUTION

This problem is essentially a calculus problem with a "flavouring" of fluid mechanics.

Referring to Fig. 41 when $t = 0$, $h = 6{\cdot}5$ m and $x = 2{\cdot}5$ m. At time t seconds h will have decreased and x will have increased. Moreover, whatever volume leaves A, the same volume enters B.

$$\therefore \qquad (6{\cdot}5 - h) \times 78 = (x - 2{\cdot}5) \times 50$$

from which, $x = 12{\cdot}64 - 1{\cdot}56h$

At time t seconds the head producing flow $= h - x$

$$= (2{\cdot}56h - 12{\cdot}64)\ \text{m}$$

For steady flow, the head lost in friction = available head, *i.e.*

$$2{\cdot}56h - 12{\cdot}63 = 4flu^2/(2gd)$$

$$= 4 \times 0{\cdot}007 \times 360u^2/(2 \times 9{\cdot}81 \times 0{\cdot}130)$$

$$u = 0{\cdot}503\sqrt{2{\cdot}56h - 12{\cdot}64} \qquad (1)$$

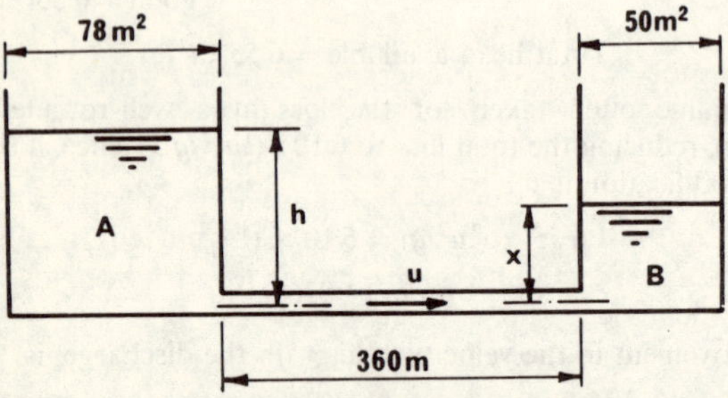

FIG. 41. Reservoir levels at time t seconds

For continuity of flow,

$$u \times \pi/4 \times 0{\cdot}130^2 = -78\,\mathrm{d}h/\mathrm{d}t$$

with a minus sign because h decreases as t increases

$$\therefore \qquad \mathrm{d}h/\mathrm{d}t = -0{\cdot}000\,170u$$

Substituting from equation (1) for u,

$$\mathrm{d}h/\mathrm{d}t = -0{\cdot}856\sqrt{2{\cdot}56h - 12{\cdot}64} \times 10^{-4}$$

The time T seconds taken for h to drop from 6·5 to 5·5 m is

$$T = \int_0^T \mathrm{d}t = -\frac{10^4}{0{\cdot}856}\int_{6\cdot5}^{5\cdot5} \frac{\mathrm{d}h}{\sqrt{2{\cdot}56h - 12{\cdot}64}}$$

$$= \frac{10^4}{0{\cdot}856} \times \frac{2}{2{\cdot}56}[\sqrt{2{\cdot}56h - 12{\cdot}64}]_{5\cdot5}^{6\cdot5}$$

$$\therefore\ T = 7310 \text{ seconds} = \underline{\underline{2{\cdot}03 \text{ hours}}}$$

ENERGY LINE, PRESSURE LINE AND HYDRAULIC GRADIENT

On a sketch of a pipeline it is sometimes convenient to draw the energy line (E.L.) and the pressure line (P.L.). Fig. 42 shows examples of these for two different flow systems.

Energy, like velocity, is not an absolute quantity, and can therefore be measured from any convenient datum. In Fig. 42 (*a*) an arbitrary datum is drawn; from this datum the energy at section *I* is $z_1 + p_1/(\rho g) + u_1^2/(2g)$ and the energy at section *II* is $z_2 + p_2/(\rho g) + u_2^2/(2g)$. In other words, the vertical distance between the energy line and the datum at any section gives the total energy there. Obviously, if a new datum were chosen the total energy at both sections *I* and *II* would be changed, but this does not matter, as the important thing, the difference in energy between sections *I* and *II*, is the same wherever the datum is chosen.

The energy line for the water in the reservoirs is always the free surface. Hence, in Fig. 42 (*a*) the water flowing from *A* into the pipe suffers an immediate loss of energy, $\frac{1}{2}u^2/(2g)$, owing to the abrupt pipe entrance. A further shock loss occurs at *P*, where the pipe section is increased, but little loss occurs at the exit into *B*, because of the diffuser. All these losses are shown by the energy line; in addition, it shows that the friction loss per unit length is greater in the smaller-bore pipe than in the larger, for friction loss per unit length is indicated by the slope of the energy line.

The pressure line is drawn everywhere at a distance $u^2/(2g)$ below the energy line, and consequently rises towards the energy line at *P*, where the flow velocity is decreased. The main significance of the pressure line is, however, that the vertical distance between the centre-line of the pipe and the pressure line is a measure to scale of the pressure energy. Thus, if stand pipes were introduced at various points along the pipe the water-level in each would rise to the pressure line. The rise of the pressure line just before the pipe exit into *B* is due to the conversion of kinetic energy into pressure energy in the diffuser. A similar rapid change of pressure (this time decreasing) is to be seen in Fig. 42 (*b*) at the well-rounded entrance to the pipe from reservoir *C*.

Figure 42 (*b*) also shows two other important features. The first of these is the exit loss of $u^2/(2g)$, which can now be interpreted as a dissipation of kinetic energy as the water is brought virtually to rest in the reservoir, *D*. The other feature is the occurrence of negative pressure head, which reaches a maximum (*i.e.* maximum negative value) at *Q*, where the pressure energy is $-p_Q/(\rho g)$. Since the driving force depends on atmospheric pressure, it would seem that the siphon would break down if this quantity exceeded 10·35 m of water (atmospheric pressure); in practice, the siphon is broken when $p_Q/(\rho g) > 7{\cdot}92$ m (approx), owing to air and other dissolved gases coming out of solution.

It should be realised that a siphon needs priming. In other words, in Fig. 42 (*b*), if the system is at rest and *Q* is above the water-level in *C*, no flow will start. If the level in *C* rises, flow will commence as soon as it passes the height of *Q*, after which flow will continue, even though the level in *C* falls below *Q*.

Notice that the pressures p_1, p_2 and p_Q are all gauge pressures, as they are calculated with regard to the free water surfaces, which are taken as being at zero pressure.

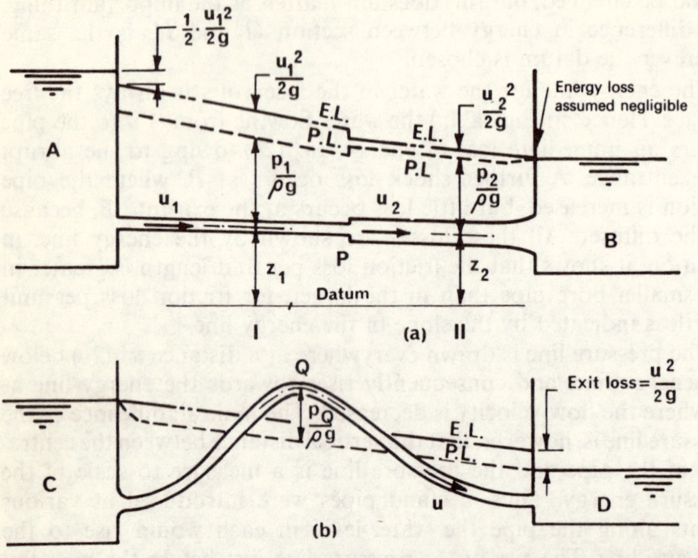

FIG. 42. (*a*) Energy line and pressure line for flow from one reservoir, *A*, to another, *B*, along a pipe which has a change of section at *P*. (*b*) Energy line and pressure line for flow from one reservoir, *C*, to another, *D*, along a pipe which forms a *siphon*

The scales of sketches, such as in Fig. 42, are usually greatly distorted, and consequently the slopes of the pipes are probably exaggerated. Because in practice these slopes are usually small, the *hydraulic gradient* is often taken as the slope of the pressure line. Some authors go further and label the pressure line as the hydraulic gradient, but this is a loose practice, for one is a *line* and the other a *slope*. Strictly, the hydraulic gradient is the loss of head per unit length of pipe, which of course, is not the same as the slope of the pressure line for a steeply sloping pipe.

EXAMPLE 27

A 300 mm diameter pipe of total length 30 m forms a siphon connecting two reservoirs in which the difference in the water-levels is 12 m. The pipe inlet is efficiently bell-mouthed, and the outlet is abrupt. The bend losses in the siphon have been observed to amount to $0{\cdot}01u^2$, and the friction coefficient $f = 0{\cdot}05$. Using the Darcy formula, $i = (4fu^2)/(2gd)$, calculate the exact discharge of the siphon (u = mean velocity, i = friction gradient, d = pipe diameter). Work in metres and seconds.

SOLUTION

The energy losses are: $u^2/(2g)$ at the abrupt exit $(4flu^2)/(2gd)$ in friction, and $0{\cdot}01u^2$ at the bend; where, in this case, $l = 30$ m. The head producing flow is 12 m, and must be equal to the sum of the losses; i.e.

$$12 = u^2/(2g).(1 + 4 \times 0{\cdot}05 \times 30/0{\cdot}3 + 0{\cdot}01 \times 2g)$$

i.e.

$$12 = u^2/(2g) \times 21{\cdot}2$$

$$\therefore u = 3{\cdot}33 \text{ m/s}$$

and discharge, $Q = 3{\cdot}33 \times \frac{\pi}{4} \times 0{\cdot}3^2 = \underline{\underline{0{\cdot}236 \text{ m}^3/\text{s.}}}$

PIPE NETWORKS AND RELATED PROBLEMS

Flow through pipes in parallel is analogous to the flow of electricity through resistances in parallel. Rules for solving the simpler problems on pipes in parallel or series can be developed, and are analogous to Kirchhoff's laws. Such problems can often be solved from first principles, however, by applying the continuity principle and a basic knowledge of energy losses in pipe-flow, as illustrated in Example 28.

EXAMPLE 28

A pipeline, with diameter 0·9 m and length 4 km, connects two fresh-water open reservoirs which have their water surfaces at elevations 75 m and 52 m above Ordnance Datum. In order to increase the rate of flow between the reservoirs by 30 per cent, it is decided to lay an additional 0·9 m diameter pipeline from the upper reservoir. The second pipeline is to lie parallel to the original pipeline and is to be connected to the latter at some suitable point. Determine the point of connection, assuming that the frictional factor, f, has the value of 0·010 for each pipeline. Minor losses may be considered to be negligible.

SOLUTION

Figure 43 represents conditions after the additional pipe has been installed.

Let h_1 = head lost between A and P, and

h_2 = head lost between P and B.

Also, let h_0 = head lost between A and B, before installing the second pipe, and

$$u_0 = \text{velocity in the pipe caused by } h_0.$$

Since the discharge is increased by 30%, $u_2 = 1{\cdot}3u_0$

i.e.

$$u_2^{\,2} = 1{\cdot}69u_0^{\,2}$$

For continuity, $u_1 = \frac{1}{2}u_2$ and, therefore,

$$u_1{}^2 = \tfrac{1}{4}u_2{}^2 = 0{\cdot}4225u_0{}^2$$

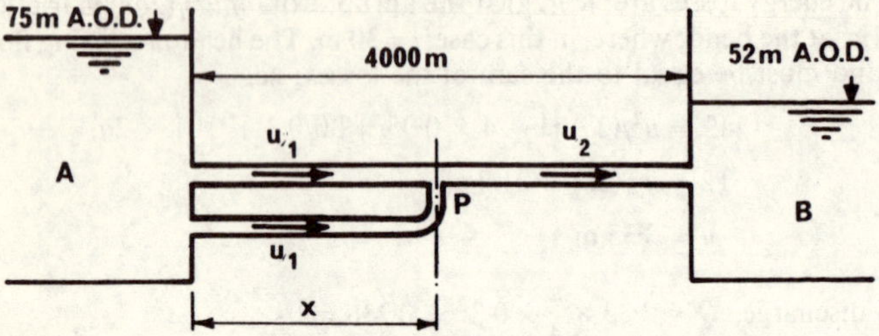

FIG. 43. *See* Example 28

In Darcy's law, $h = (4flu^2)/(2gd)$, f, g and d are the same for both pipes. Hence,

$$u_0{}^2 \propto h_0/l_0 = i_0, \text{ say}$$

$$u_1{}^2 \propto h_1/l_1 = i_1$$

and $$u_2{}^2 \propto h_2/l_2 = i_2$$

From the conditions shown on Fig. 43, $i_0 = 23/4000 = 0{\cdot}005\,75$ and, since $u_2{}^2 = 1{\cdot}69u_0{}^2$,

$$i_2 = 1{\cdot}69i_0 = 0{\cdot}009\,72$$

and, similarly, $i_1 = \frac{1}{4}i_2 = 0{\cdot}002\,43$

If the distance from A to P is x m,

$$h_1 = i_1x$$

i.e. $$h_1 = 0{\cdot}002\,43\,x$$

and $$h_2 = 0{\cdot}009\,72(4000 - x)$$

Since, $$h_1 + h_2 = 23 \text{ m}$$

$$0{\cdot}002\,43x + 0{\cdot}009\,72(4000 - x) = 23$$

Hence length of pipe, $x = 2180$ m = 2·18 km

For more complex pipe networks, iterative methods have to be adopted. Example 29 below provides a convenient introduction to these methods because it can be solved by an "exact" method—*i.e.* the first solution—and also because it lends itself to solution by successive approximations: *see* p. 97 for the alternative solution.

EXAMPLE 29

Two water-supply reservoirs, at elevations 170 and 192 m O.D. (Ordnance Datum), are connected by 0·3 m diameter pipes, 3 km long, to a 0·38 m diameter main, also 3 km long, which feeds a service reservoir at 85 m O.D. The roughness coefficient, n, of all pipes is the same. Using Manning's formula, $u = (1{\cdot}00/n)m^{\frac{2}{3}}i^{\frac{1}{2}}$, determine the flow into the service reservoir. (u=mean velocity; n=0·014; m=hydraulic radius; i=hydraulic energy gradient.)

SOLUTION

See Fig. 44.

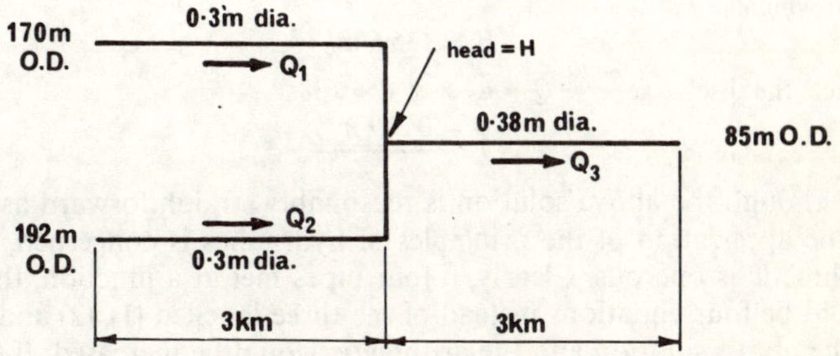

FIG. 44 *See* Example 29

For a round pipe, the hydraulic mean radius is:

$$m = \frac{\pi d^2}{4} \div \pi d = d/4, \text{ and}$$

$$\therefore \quad u = \frac{1}{n}\left(\frac{d}{4}\right)^{2/3} \cdot \quad i^{1/2} = k' d^{2/3} i^{1/2} \quad \text{where} \quad k' = \frac{1}{0{\cdot}014 \times 4^{2/3}}$$

Referring to the pipes in Fig. 44 by the suffices 1, 2 and 3, respectively, $d_1 = d_2 = 0{\cdot}3$ m and $d_3 = 0{\cdot}38$ m. Also:

$$i_1 = \frac{170 - H}{3000} \qquad u_1 = k \times 0{\cdot}3^{2/3}\sqrt{170 - H} \tag{1}$$

$$i_2 = \frac{192 - H}{3000} \qquad \therefore \quad u_2 = k \times 0{\cdot}3^{2/3}\sqrt{192 - H} \tag{2}$$

$$i_3 = \frac{H - 85}{3000} \qquad \therefore \quad u_3 = k \times 0{\cdot}38^{2/3}\sqrt{H - 85} \tag{3}$$

where H is the total head at the junction of the three pipes, (*see* Fig. 44), and $k = k'/\sqrt{3000} = 0{\cdot}518$.

For continuity,

$$u_3 \times \pi/4 \times 0{\cdot}38^2 = \pi/4 \times 0{\cdot}3^2(u_1 + u_2)$$

from which,

$$u_3 = 0{\cdot}623(u_1 + u_2)$$

Substituting for u_1, u_2, u_3 from equations (1), (2), (3):

$$\sqrt{H-85}=\left(\frac{0{\cdot}30}{0{\cdot}38}\right)^{2/3}\times 0{\cdot}623(\sqrt{170-H}+\sqrt{192-H})$$

$$H-85=0{\cdot}283(362-2H+2\sqrt{(170-H)(192-H)})$$

$$\therefore \qquad 1{\cdot}567H-187{\cdot}6=0{\cdot}566\sqrt{32\,640-362H+H^2}$$

Squaring both sides and collecting like terms leads to:

$$6{\cdot}640H^2-1467H+76\,882=0$$

from which:

$$H=135{\cdot}6 \text{ m}$$

Hence, the discharge: $Q = u_3 \times \pi/4 \times 0{\cdot}38^2$

$\therefore$ $\underline{\underline{Q = 0{\cdot}219\ m^3}}$

Although the above solution is reasonably straightforward as far as the application of the principles of hydraulics is concerned, the arithmetic is onerous. Clearly, if four pipes met at a junction, there would be four equations instead of the three labelled (1), (2) and (3) in the above solution and the arithmetic would be increased. If this were just one junction of many in a network, the work involved in arriving at a solution would become overwhelming. For this reason alternative methods of solution using a process of successive approximations have been devised and lend themselves to programming on a computer. The first of these to be described is based on the following theory and is particularly applicable to a junction.

Junction of a network

The Chézy, Darcy and Manning formulae, for example, are all of the form $h = kQ^n$ in which the exponent n is often taken (for convenience and simplicity) as 2 although in more accurate work it is sometimes given a non-integral value. h is the head causing flow, Q is the rate of flow in m^3/s and k is a constant which depends on the pipe diameter, length, shape and roughness, which may very well vary from one pipe to another meeting at the same junction.

Assuming, therefore, that the flow in each pipe at the junction obeys the law $h = kQ^n$, it follows that:

$$\mathrm{d}h = nkQ^{n-1}\mathrm{d}Q = \frac{nh}{Q}\mathrm{d}Q$$

i.e.

$$\mathrm{d}Q = \frac{Q}{nh}\mathrm{d}h$$

For continuity, the total flow into the junction must equal the total flow away from the junction. The flow *in* may be taken as positive and the flow *out* taken as negative. If the total head at the junction is *guessed* and the discharge calculated for each pipe meeting there, the guess will be correct only if the algebraic sum of the Q-values is zero, *i.e.*, if $Q_1 + Q_2 + Q_3 + \ldots = 0$ where Q_1, Q_2, Q_3 *etc.* are the discharges in each pipe. Usually, the guess of the total head will not give an exact balance (it may be well out). Then the sum of the calculated values will be:

$$Q_1 + Q_2 + Q_3 + \ldots = \Delta Q$$

Since the right-hand side should be 0, each value of Q has to be changed by an amount dQ_1, dQ_2, dQ_3 *etc.* and, for continuity, these must satisfy the equation:

$$dQ_1 + dQ_2 + dQ_3 + \ldots = \Delta Q$$

Hence
$$\frac{dh}{n}\left(\frac{Q_1}{h_1} + \frac{Q_2}{h_2} + \frac{Q_3}{h_3} + \ldots\right) = \Delta Q$$

i.e.
$$dh = \frac{n\Delta Q}{\Sigma(Q/h)}$$

dh takes the sign of ΔQ and can, therefore, be positive or negative. The Q/h terms are all positive.

If the value for total head was initially assumed to be H, a better solution will be obtained by using $H + dh$ and recalculating Q. By successive repetitions of this process a point will be reached at which the adjustment dh is negligible—and then a *head balance* is said to be obtained.

The method is illustrated in the following alternative solution to Example 29.

ALTERNATIVE SOLUTION

Reference should be made to Fig. 44 and the original solution on p. 94. The following expressions for Q are calculated from

$$Q = \left(\frac{\pi}{4}\right) d^2 \times \left(\frac{1}{n}\right)\left(\frac{d}{4}\right)^{2/3} \times \left(\frac{\sqrt{h}}{\sqrt{3000}}\right)$$

$$Q_{1,2} = \frac{\pi}{4} \times 0{\cdot}3^2 \times \frac{1}{0{\cdot}014} \times \left(\frac{0{\cdot}3}{4}\right)^{2/3} \times \frac{\sqrt{h_{1,2}}}{\sqrt{3000}} = 0{\cdot}0164\sqrt{h_{1,2}}$$

$$Q_3 = \frac{\pi}{4} \times 0{\cdot}38^2 \times \frac{1}{0{\cdot}014} \times \left(\frac{0{\cdot}38}{4}\right)^{2/3} \times \frac{\sqrt{h_3}}{\sqrt{3000}} = 0{\cdot}0308\sqrt{h_3}$$

Guess $H = 140$ m

Pipe	h	$\sqrt{h}$	Q	Q/h
1	30	5·48	0·090	0·003
2	52	7·21	0·118	0·002
			0·208	
3	55	7·46	−0·228	0·004
		$\Delta Q =$	−0·020	$0{\cdot}009 = \Sigma(Q/h)$

$$\therefore \qquad \mathrm{d}h = \frac{2\Delta Q}{\Sigma(Q/h)} = -\frac{0{\cdot}04}{0{\cdot}009} = -4{\cdot}4$$

Second approximation, $H = 140 - 4{\cdot}4 = 135{\cdot}6$

Pipe	h	$\sqrt{h}$	Q	Q/h
1	34·4	5·865	0·096	0·003
2	56·4	7·510	0·123	0·002
			0·219	
3	50·6	7·113	−0·219	0·004
			$\Delta Q = 0$	$0{\cdot}009 = \Sigma(Q/h)$

$\therefore$ d$h = 0$ and, as in the first solution, $H = 135{\cdot}6$ m, giving

$$Q = 0{\cdot}219 \text{ m}^3/\text{s}$$

Before proceeding, the reader is advised to repeat the above calculation starting with an inferior initial guess of, say, $H = 150$ m to satisfy himself that the method does converge fairly quickly to the right answer regardless of how accurate (or lucky) the first guess is.

Ring main

The procedure for flow computations in a ring main is similar to that described for pipes meeting at a junction. The main difference is that, in the case of a ring main, the result is usually obtained by seeking a *quantity balance* rather than a head balance. Simple diagrams of a ring main are shown in Fig. 45 and Fig. 46. The theory upon which the method is based is very similar to that given on p. 96.

Again it is assumed that flow in each pipe of the ring obeys the law $h = kQ^n$, the symbols being as before. In this method, however, the first step is to assume a value for the flow in one of the pipes and derive the values of the flow in all the other pipes from it. The corresponding head loss in each of the pipes can then be calculated from $h = kQ^n$ and, if the guess happened to be correct, the sum of the heads taken round the ring would be zero. For the purpose of summing them, the sign convention used is that the head is positive for a pipe in which the flow is in the clockwise direction round the ring. In practice, the sum will not normally be equal to zero, but to a closing error Δh.

i.e. $$\Delta h = h_1 + h_2 + h_3 + \ldots = \Sigma h$$

Since $h = kQ^n$, $\mathrm{d}h = nkQ^{n-1}\mathrm{d}Q = (nh/Q)\mathrm{d}Q$

The closing error is distributed by making adjustments $\mathrm{d}h_1$, $\mathrm{d}h_2$ *etc.* to each pipe in turn and:

$$\Delta h = \mathrm{d}h_1 + \mathrm{d}h_2 + \mathrm{d}h_3 + \ldots$$
$$= (nh_1/Q_1)\mathrm{d}Q_1 + (nh_2/Q_2)\mathrm{d}Q_2 + \ldots$$

Since the flows were initially balanced, for continuity to be preserved:

$$\mathrm{d}Q_1 = \mathrm{d}Q_2 = \mathrm{d}Q_3 = \ldots = \Delta Q$$

$$\therefore \qquad \Delta h = n\Delta Q\Sigma(h/Q)$$

$\therefore$ the adjustment to be made is

$$Q = -\frac{\Delta h}{n\Sigma(h/Q)}$$

EXAMPLE 30

In a simple ring main ABC water is supplied at C at the rate of 0·15 m^3/s and drawn off at A and B at the rates of 0·13 m^3/s and 0·02 m^3/s respectively. If the loss of head in each pipe obeys the law $h = KQ^2$, determine the flow rates in each pipe for the values of K given on Fig. 45 (*a*).

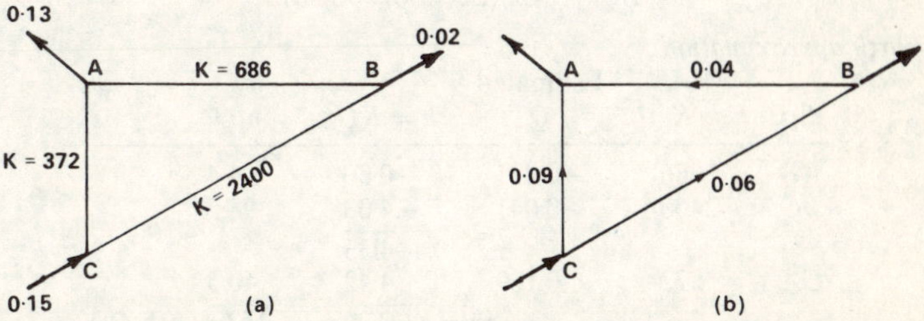

FIG. 45. *See* Example 30: (*a*) the pipe network and (*b*) the first guess of Q values

SOLUTION

Assume the flow in CA to be 0·09 m^3/s, clockwise. Then flow in AB is

$$Q_{AB} = 0{\cdot}09 - 0{\cdot}13 = -0{\cdot}04 \text{ m}^3\text{/s, and}$$

$$Q_{BC} = -0{\cdot}04 - 0{\cdot}02 = -0{\cdot}06 \text{ m}^3\text{/s.}$$

First approximation

Pipe	K	Estimated Q	$h = KQ^2$	h/Q
AB	686	$-0{\cdot}04$	$-1{\cdot}10$	27·5
BC	2400	$-0{\cdot}06$	$-8{\cdot}64$	144·0
			$-9{\cdot}74$	
CA	372	$+0{\cdot}09$	$+3{\cdot}01$	33·5
			$\Delta h = -6{\cdot}73$	$205 = \Sigma(h/Q)$

$$\Delta Q = +6{\cdot}73/(2 \times 205) = 0{\cdot}016 \text{ (Clockwise, since positive.)}$$

Second approximation

Pipe	K	Estimated Q	$h = KQ^2$	h/Q
AB	686	$-0{\cdot}024$	$-0{\cdot}39$	16
BC	2400	$-0{\cdot}044$	$-4{\cdot}65$	106
			$-5{\cdot}04$	
CA	372	$+0{\cdot}106$	$+4{\cdot}18$	39
			$\Delta h = -0{\cdot}86$	$161 = \Sigma(h/Q)$

$$\Delta Q = 0{\cdot}86/(2 \times 161) = 0{\cdot}002$$

Third approximation

Pipe	K	Estimated Q	$h = KQ^2$	h/Q
AB	686	$-0{\cdot}022$	$-0{\cdot}33$	15
BC	2400	$-0{\cdot}042$	$-4{\cdot}23$	101
			$-4{\cdot}56$	
CA	372	$0{\cdot}108$	$4{\cdot}34$	40
			$\Delta h = -0{\cdot}22$	$156 = \Sigma(h/Q)$

$$\Delta Q = 0{\cdot}22/(2 \times 156) = 0{\cdot}001$$

Fourth approximation

Pipe	K	Estimated Q	$h = KQ^2$	h/Q
AB	686	$-0{\cdot}021$	$-0{\cdot}30$	14·4
BC	2400	$-0{\cdot}041$	$-4{\cdot}03$	98·3
			$-4{\cdot}33$	
CA	372	$0{\cdot}109$	$4{\cdot}42$	40·5
			$\Delta h = +0{\cdot}09$	$153 = \Sigma(h/Q)$

$$\Delta Q = -0{\cdot}09/(2 \times 153) = -0{\cdot}0003$$

The required flow rates are, therefore:

$$\underline{\underline{Q = 0{\cdot}021 \text{ m}^3/\text{s from } B \text{ to } A}}$$

$$\underline{\underline{Q = 0{\cdot}041 \text{ m}^3/\text{s from } C \text{ to } B}}$$

$$\underline{\underline{Q = 0{\cdot}109 \text{ m}^3/\text{s from } C \text{ to } A}}$$

The above solution, in practice, would not have been taken past the third approximation and was done in the text only to demonstrate the change of sign in ΔQ. In practice there are many imponderables and the problems do not, therefore, warrant a high degree of computational accuracy.

Example 31 deals with a more complicated pipe system in which there are two ring mains with one pipe common to both. The method used in the last solution is again employed but has to be applied to each ring in turn and, of course, adjustment of the flow in one circuit tends to upset a balance in the other. Repeated applications of the procedure, however, soon converge to a final solution of acceptable accuracy.

EXAMPLE 31

An existing water distribution system for a small housing estate is shown in Fig. 46. The mains are to be renewed in cast iron maintaining the same inflow at C and the same outflows at A, B, and D. Details of the new pipes are as follows:

pipe:	AB	BC	CD	DA	BD
diameter (mm):	300	300	300	300	250
length (km):	1·0	0·5	1·0	0·5	1·3

Determine the flow rates in each pipe to 2 decimal places using the Moody Diagram (*see* Fig. 84) and assuming the head loss due to friction is given by

$h = \dfrac{fLQ^2}{3d^5}$ and the kinematic viscosity of water is $\nu = 0{\cdot}01\ \text{cm}^2/\text{s}$.

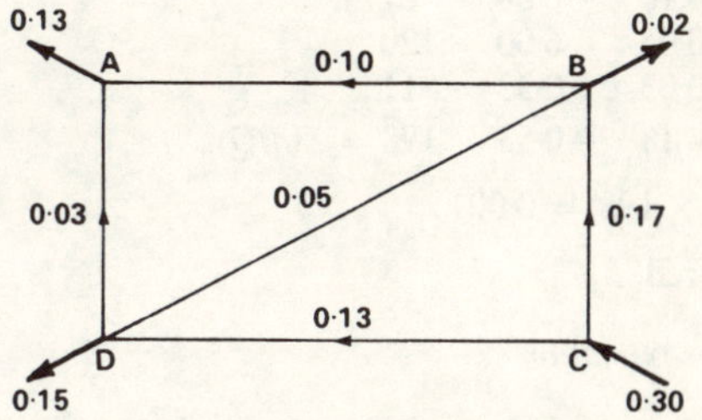

FIG. 46 *See* Example 31

SOLUTION

The Moody diagram is shown in Fig. 84 and is explained in Chapter 11 on p. 187. Basically, it is a graph showing how the friction factor f varies with the Reynolds number for a range of pipes which have different relative roughnesses k/d. k is the equivalent size in mm of projections on the inside of the pipe wall and, for cast iron k is 0·3 and, therefore, for four of the pipes, $k/d = 0{\cdot}3/300 = 0{\cdot}001$ whilst for BD, $k/d = 0{\cdot}0012$.

For pipe AB, $Q = 0{\cdot}1$ and, therefore, $u = 0{\cdot}1 \div \frac{1}{4}\pi d^2 = 1{\cdot}41$ m/s

$$R_e = ud/\nu = 1{\cdot}41 \times 0{\cdot}3 \div 10^{-6} = 4{\cdot}24 \times 10^5$$

For pipe AB, $K = fL/(3d^5) = 0{\cdot}0050 \times 1000/(3 \times 0{\cdot}3^5)$

$$= 686$$

Similar calculations for the other pipes give the following values:

Pipe	Q	R_e	k/d	f	L	K
AB	0·10	$4{\cdot}24 \times 10^5$	0·001	0·0050	1000	686
BC	0·17	$7{\cdot}22 \times 10^5$	0·001	0·0050	500	343
CD	0·13	$5{\cdot}52 \times 10^5$	0·001	0·0050	1000	686
DA	0·03	$1{\cdot}27 \times 10^5$	0·001	0·0055	500	372
BD	0·05	$2{\cdot}55 \times 10^5$	0·0012	0·0054	1300	2400

The above values for Q were the values existing in the old installation but are sufficiently accurate for the determination of the f-values.

In the following computation the two rings ABD and BCD are taken separately and it should be noted how adjustments to one affect the other. It should also be noticed that the flow in pipe BD counts as clockwise for ring ABD but as anti-clockwise for ring BCD.

ABD—First approximation

Pipe	K	Q	h	h/Q
AB	686	−0·10	−6·86	67
BD	2400	0·05	6·00	120
DA	372	0·03	0·33	11
			$\Delta h = -0{\cdot}53$	$198 = \Sigma(h/Q)$

$\therefore \Delta Q = 0{\cdot}53/(2 \times 198) = 0{\cdot}001$

and can be ignored.

BCD—First approximation

Pipe	K	Q	h	h/Q
BC	343	−0·17	−9·91	58
CD	686	0·13	11·59	89
DB	2400	−0·05	−6·00	120
			$\Delta h = -4{\cdot}32$	$267 = \Sigma(h/Q)$

$\therefore \Delta Q = 4{\cdot}32/(2 \times 267) = +0{\cdot}01$

Hence Q_{BD} changed and ABD has to be reconsidered.

ABD—Second approximation

Pipe	K	Q	h	h/Q
AB	686	−0·10	−6·86	67
BD	2400	0·04	3·84	96
DA	372	0·03	0·33	11
			$\Delta h = -2{\cdot}67$	$174 = \Sigma(h/Q)$

$\therefore\ \Delta Q = 2{\cdot}67/(2 \times 174) = 0{\cdot}01$

BCD—Second approximation

Pipe	K	Q	h	h/Q
BC	343	−0·16	−8·80	55
CD	686	0·14	13·40	96
DB	2400	−0·05	−6·00	120
			$\Delta h = -1{\cdot}4$	$271 = \Sigma(h/Q)$

$\therefore\ \Delta Q = 1{\cdot}4/(2 \times 271) = 0{\cdot}003$

The distribution of flow at each stage is shown in Fig. 47 (*a*), (*b*), (*c*) and (*d*), the final flow correct to two decimal places being as shown by Fig. 47 (*d*).

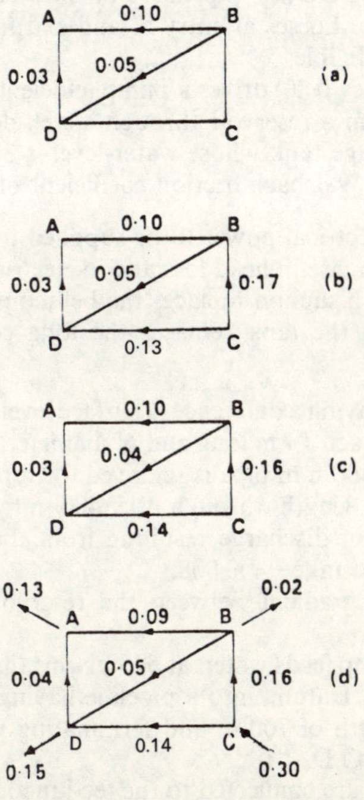

FIG. 47. *See* Example 31: (*a*) *ABD*—first approximation; (*b*) *BCD*—first approximation; (*c*) *ABD*—second approximation; (*d*) *BCD*—second approximation

QUESTIONS

1. What assumptions are made in the derivation of Bernoulli's theorem from the principle of the conservation of energy?

A constant-speed pump is capable of delivering 28 litre/s through a 100 mm diameter pipe 120 m long, from a lake to a tank 6 m above the lake level. Calculate the pipe diameter required if the same pump is to deliver 28 litre/s to the same tank raised to 12 m above the lake level.

Use Manning's formula: $u=(1{\cdot}00/n).m^{\frac{2}{3}}i^{\frac{1}{2}}$
where u is the velocity in m/s; m the hydraulic mean depth in m; i is the slope; and n is the roughness coefficient, which may be assumed to be 0·012.

Would you in practice increase the diameter of the whole length of the pipe?

2. Two open water-reservoirs are connected by 800 m of 0·6 m internal-diameter pipeline and the resulting gravitational flow, caused by the difference in levels of the reservoirs, is 0·5 m^3/s. For the pipeline, $f=0{\cdot}010$ in the Darcy equation.

Determine the total flow between the reservoirs when an additional pipeline, 0·6 m internal diameter and 550 m in length, has been laid from the higher reservoir, parallel to the first pipeline and connected to it at a point 550 m from the inlet to the first pipeline. For the second pipeline, $f = 0{\cdot}005$ in the Darcy equation. Losses at entry to and exit from the pipelines may be assumed to be negligible.

3. A motor (efficiency 0·80) drives a pump (efficiency 0·65) which delivers 12 litre/s of water from a reservoir through a well-designed 100 mm pipe, 850 m long, to a storage tank whose water-level is 30 m above that of the reservoir. The Darcy–Weisbach friction coefficient of the pipe is $f = 0{\cdot}015$.

(*a*) Calculate the electrical power to be supplied to the motor.
(*b*) Estimate the increase in head loss and in electrical power if an abrupt inlet to the pump suction replaces the bellmouth and an abrupt discharge end into the tank replaces the long conical diffuser in the original design.

4. Two reservoirs having a difference of surface level of 30 m are connected by two parallel pipes each 1 km long and of diameter 300 mm and 400 mm. A length of 200 m of 300 mm pipe is removed for repair and the total flow is diverted over this length through the 400 mm pipe. Calculate the percentage reduction in discharge resulting from the diversion. Consider friction losses only and take $f = 0{\cdot}008$.

Plot the hydraulic gradient between the reservoirs when the flow is diverted.

5. A centrifugal pump feeds water, at 545 kN/m^2 (gauge) and at elevation 58 m above Ordnance Datum, into a pipeline having an internal diameter of 500 mm and a length of 700 m and terminating with a tee-junction at elevation 58 m above O.D.

Two other pipelines are connected to the tee-junction. One of these pipelines has an internal diameter of 400 mm and a length of 260 m, and is connected to a reservoir which has its free water surface at elevation 100 m above O.D. The third pipeline has a diameter of 300 mm and a length of 550 m, and discharges water into a distribution network at a point where the elevation is 67 m above O.D. and the gauge-pressure is 194 kN/m^2.

With sample calculations (and assuming $f = 0{\cdot}010$ for each pipeline), demonstrate a method of successive corrections by means of which the quantity flowing in the three pipelines, and the pressure at the tee-junction, can ultimately be determined.

6. Details of part of a pipeline are given in Fig. 48. When the rate of flow from *A* to *C* is 1·5 litres per second, the water level in the piezometer at *C* is 1·0 m above the centre-line of the pipe.

(*a*) Determine the height of the water level in the piezometer at *A* if the Darcy friction factor for the pipe between *A* and *C* is 0·003. Take the loss at the sudden contraction as $0{\cdot}5u_2^2/(2g)$.

(*b*) Make a carefully dimensioned sketch of the hydraulic gradient and the total energy line. Indicate clearly all energy losses..

7. One mesh in a pipe network is equivalent to a hexagon *ABCDEFA* with all six pipes having the same diameter, length and friction coefficient. At peak inflows and outflows, as shown in column I, all flows are turbulent. At minimum inflows and outflows, as shown in column II, all flows are laminar (with the Darcy–Weisbach $f = 16/R_e$). Calculate the distribution of discharges in the two cases.

If you use an iterative method, one iteration will suffice to show your knowledge of the technique.

Node	I		II	
	Inflow	*Outflow*	*Inflow*	*Outflow*
A	1000		10	
B		400		4
C		300		3
D	500		5	
E		300		3
F		500		5

(All figures in m^3/hr.)

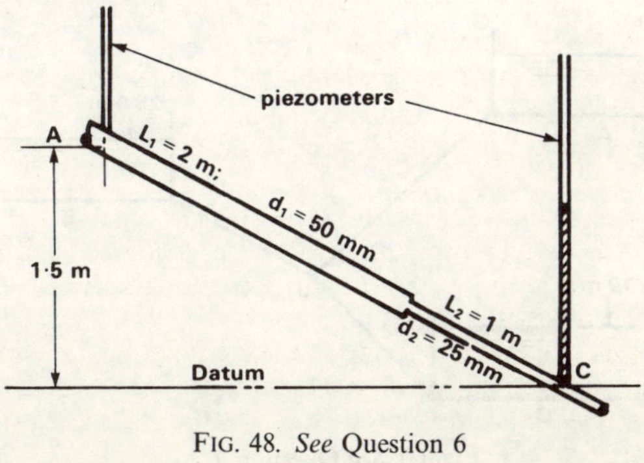

FIG. 48. *See* Question 6

8. One mesh, *ABCDA*, of a water-supply pipe network has a design inflow of 0·25 m^3/s at *A* and design demands of 0·05, 0·10, and 0·10 m^3/s at *B*, *C* and *D*. The demands at *B* and *D* are stable but there is an increase in demand at *C* amounting to 0·01 m^3/s per year, with a corresponding increase of inflow at *A*. The initial increase can be met by the construction of a main along the diagonal *BED* before a "doubler" main *AEC* has to be constructed. The pressure at *A* cannot be increased, nor may that at *C* be reduced. All mains, old and new, may be assumed to have the same friction coefficient $f = 0{\cdot}005$ in the Darcy–Weisbach formula.

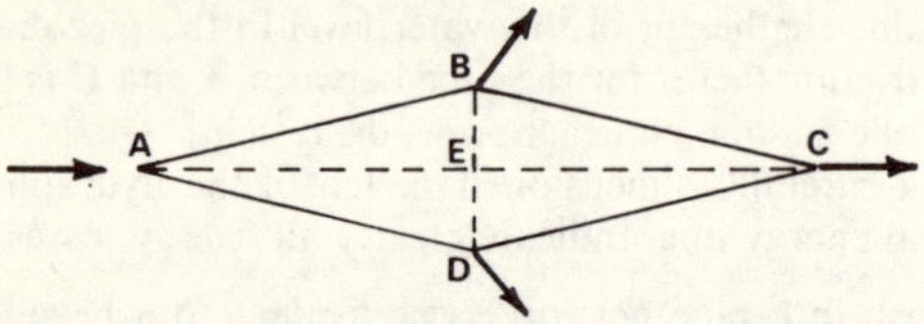

FIG. 49. *See* Question 8

		Length (m)	*Diameter* (mm)
Main existing	*AB*	1000	150
	BC	1000	300
	AD	1000	450
	DC	1000	150
Proposed	*BE*	100	300
	ED	100	300
	AE	500	450
	EC	500	300

Without attempting an accurate solution, derive the scheme of algebraic equations for solution, using numerical methods, which would permit pressures and flow rates at *C* to be calculated (*a*) for the original mesh,

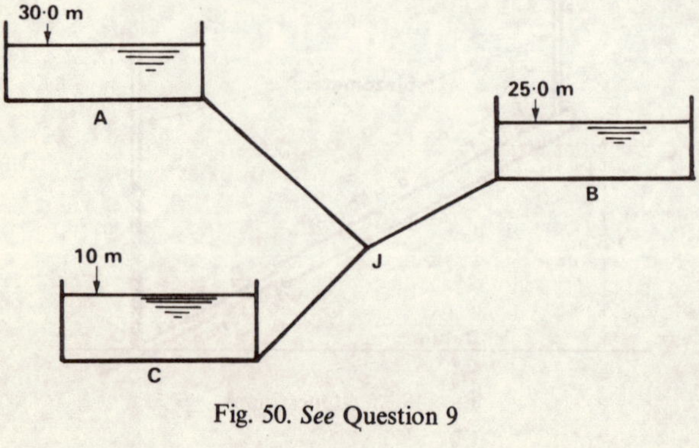

Fig. 50. *See* Question 9

(*b*) for the mesh with the diagonal *BED*, and (*c*) for the mesh with both diagonals *BED* and *AEC*.

9. Distinguish between the methods of quantity and head balance when investigating pipe networks.

Three reservoirs are connected to a common junction as shown in Fig. 50. Assuming the Darcy–Weisbach expression for energy loss to be valid, determine the flow in each of the connecting pipes, if it is known that the piezometric head at the junction is approximately 23 m. Two iterations only are required.

Pipe	*Diameter* (mm)	*Length* (km)	*f*
AJ	1·0	2	0·005
BJ	1·5	5	0·006
CJ	1·0	1	0·004

CHAPTER 8

POTENTIAL FLOW

Vorticity is one of the words used to describe the type of flow in which every fluid element is turning about an axis within itself; this may also be described as *rotational* flow. It is dealt with again later in the chapter but only to give emphasis, by contrast, to *irrotational* flow which is the subject of this chapter.

The description, *irrotational flow*, implies that there are no rotating fluid elements. Most of the flow situations described in this book are irrotational including, perhaps surprisingly, flow round a bend in an open channel and the flow in a *free vortex*. In both of these the flow path is distinctly curved (circular in the case of the free vortex) but the fluid elements follow these paths by deforming in shape and not by rotating about an inner axis. Moreover, as advanced mathematics using vector notations is required for the analysis of irrotational flow of *compressible* fluids, this chapter is concerned only with steady, inviscid flow of incompressible fluids.

PROPERTIES OF STEADY, IRROTATIONAL FLOW OF INCOMPRESSIBLE, INVISCID FLUIDS

1. Fluid particles follow well-defined paths, called *streamlines*, which do not vary with time. Streamlines cannot cross one another and, although the streamlines cannot be said to be strictly parallel to one another, they often diverge from the parallel state only gradually. Where the flow is contained by impervious boundaries, they (*i.e.* the boundaries) constitute the limiting streamlines. In any flow pattern there are virtually an infinite number of streamlines but, when the technique of sketching streamlines is used in problem solving, only a few convenient ones are selected and drawn..

2. A *stream function* ψ may be defined as having the property that ψ is a constant on any streamline. Usually the streamlines drawn are at convenient values of ψ, *e.g.* $\psi = 0, 1, 2, 3, \ldots$

3. A *velocity potential* ϕ exists. If lines $\phi =$ *constant* are plotted they form a series of curves which intersect all the streamlines at right-angles, *see* Fig. 51; *i.e.* the two sets of curves $\psi =$ constant and $\phi =$ constant are orthogonal. The lines $\phi =$ constant are called *equipotentials*. Since a velocity potential always exists for irrotational flow (and cannot exist for rotational flow), irrotational flow is also called *potential flow*.

4. In a flow problem, if the boundaries are plotted to scale, one way of arriving at a unique solution of the problem is to sketch the streamlines and equipotentials and adjust them by a series of successive approximations until they satisfy the conditions of ϕ and ψ. The resulting drawing is called a *flow net* and an example is given in Fig. 51. As the movement of most fluids through permeable beds behaves in accordance with potential flow theory, many seepage problems encountered in soil mechanics and filtration engineering lend themselves to the flow net approach. For a fuller treatment of this method, therefore, reference should be made to books on these subjects, for example, the book in this series of textbooks called *Soil Mechanics* by M. J. Smith.

Suffice it to say here that the ϕ- and ψ-lines are chosen so that they form figures which approximate to squares and that the art of drawing accurate flow nets lies in being able to distinguish between a good"square" and an unsatisfactory "square". In every flow net a few *singular fields* will occur which will clearly not approximate to squares but these are satisfactory if they subdivide into four smaller fields which are more nearly square.

EXAMPLE 32

A sheet pile wall is driven to a depth of 4 m into a permeable soil stratum, 9 m thick, which overlies impermeable bedrock. The difference in water level between the two sides of the sheet pile wall is H.

(*a*) Draw a flow net for the seepage through the permeable layer.
(*b*) Derive an expression for the rate of flow from the high pressure side of the sheet piling to the low pressure side, given that the Darcy formula for flow through a permeable material is $Q = kAi$ where Q(m^3/s) is the rate of flow across an area A under a hydraulic gradient i and k is the coefficient of permeability in m/s.
(*c*) Evaluate Q if $k = 10^{-5}$ m/s and $H = 2$m.

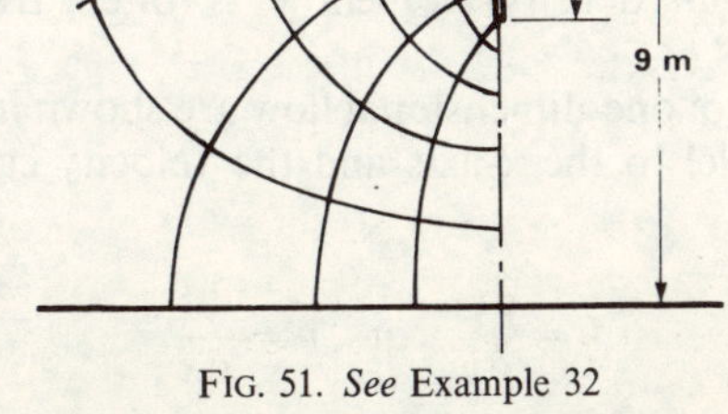

FIG. 51. *See* Example 32

SOLUTION

(*a*) The flow net is drawn to scale in Fig. 51.

(*b*) There can be no flow across a streamline which, therefore, behaves like an impervious boundary in that the principle of continuity applies throughout the space bounded by any two streamlines.

Making the fields approximate to squares ensures that the pressure drop is the same across each field. For example, in Fig. 51 the potential at the upstream ground surface is H and, as there are ten pressure drops in a complete flow path (to the downstream ground surface), the pressure will drop by $\frac{H}{10}$ across each square and the equipotentials will be, therefore, $0{\cdot}9H$, $0{\cdot}8H$, $0{\cdot}7H$, etc. In general, if there are N_e equipotential drops, the head lost across each square is $\frac{H}{N_e}$. If the length of side of a "square" is l, the hydraulic gradient is $i = \frac{H}{N_e l}$ and the cross-sectional area of one flow path is $1 \times l$ per metre of sheet pile wall.

$$\therefore \text{ flow between two adjacent streamlines} = kiA = \frac{kH}{N_e}$$

If the number of flow paths $= N_f$

total flow is:
$$Q = kH\frac{N_f}{N_e}$$

(*c*) When $k=10^{-5}$, $H=2$ and from Fig. 51. $N_f/N_e=\frac{5}{10}$,

$$Q = 10^{-5} \times 2 \times \frac{5}{10} = 10^{-5} \text{ m}^3\text{/s per metre of wall}$$

MATHEMATICAL DEFINITIONS OF ϕ AND ψ

One-dimensional flow

In one-dimensional flow, changes of velocity, pressure and other fluid properties, transverse to the main direction of flow, are ignored. In practice, this usually implies that at a cross-section the fluid properties are expressed in terms of their mean values (*e.g. mean* velocity and *mean* density). Pipeflow is often treated as one-dimensional.

Two examples of one-dimensional flow are shown in Fig. 52. In (*a*) the flow is parallel to the x-axis and the velocity can be obtained from either:

$$u = \frac{d\psi}{dy} \quad \text{or} \quad u = \frac{d\phi}{dx}$$

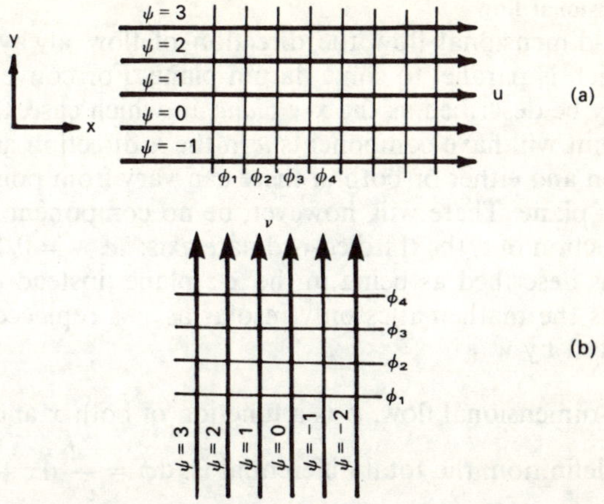

FIG. 52. (*a*) Flow parallel to the *x*-axis. (*b*) Flow parallel to the *y*-axis

The convention used for determining the sign of the stream function is that the flow is seen to cross from left to right when looking in the direction of ψ increasing. The convention used here for ϕ is that it increases in the direction of flow, *i.e.* $\phi_1 < \phi_2 < \phi_3 < \ldots$

In (*b*) the flow is parallel to the *y*-axis and, using the above sign conventions:

$$v=\frac{\mathrm{d}\psi}{\mathrm{d}x}, \text{ negative because } \psi \text{ increases in the}$$

direction of x decreasing; and

$$v=\frac{\mathrm{d}\phi}{\mathrm{d}y}.$$

It should be noted that these conventions are arbitrary and, in fact, the opposite signs of either or both may be used instead (*see* Example 33). A change of convention leads to a change of sign of u and v but does not change the final result.

It should also be noted that the important thing about ψ is, strictly, the amount by which it changes from one streamline to the next. In Fig. 52 (*a*), for example, instead of using $\psi = -1, 0, 1, 2, 3, \ldots$, the values $\psi = 8, 9, 10, 11, 12, \ldots$ could have been used instead without affecting the flow. In problems such as Question 2 at the end of this chapter, however, it is usually convenient to choose ψ so that $\psi = 0$ when $x = 0$ and $\psi = 0$ when $\theta = 0$. Similarly, for ϕ, the important feature is, strictly, the amount by which it changes from one equipotential to the next although, in its case, appropriate values are usually suggested by external conditions, as in Example 33.

Two-dimensional flow

In two-dimensional flow the direction of flow always lies in a plane which is parallel to some datum plane. For convenience, the plane may be described as the *x-y plane*, in which case, the velocity at any point will have components u in the x-direction and v in the y-direction and either or both of these can vary from point to point in the x-y plane. There will, however, be no component of velocity in the direction of z, the third co-ordinate axis; *i.e.* $w = 0$. Sometimes the flow is described as being in the x-z plane (instead of x-y) but this affects the mathematics only insofar as y is replaced by z and v is replaced by w.

In two-dimensional flow, ϕ is a function of both x and y, and so is ψ. By definition, the total differential is: $d\phi = \frac{\partial \phi}{\partial x}dx + \frac{\partial \phi}{\partial y}dy$

$$\therefore \qquad d\phi = u\,dx + v\,dy$$

where $u = \frac{\partial \phi}{\partial x}$ and $v = \frac{\partial \phi}{\partial y}$.

Similarly, $$d\psi = -v\,dx + u\,dy$$

where $-v = \frac{\partial \psi}{\partial x}$ and $u = \frac{\partial \psi}{\partial y}$.

Partial derivatives are used here in the sense that $\frac{\partial}{\partial x}$ implies that y is being kept constant and $\frac{\partial}{\partial y}$ implies x is being kept constant.

Example 33 illustrates the use of these relationships in plotting streamlines and equipotentials. First, however, this is a convenient place in the text to consider briefly:

(*a*) The mathematical condition for the flow to be irrotational is

$$\frac{\partial u}{\partial y} = \frac{\partial v}{\partial x}$$

and this is satisfied by the above equations for ϕ because

$$\frac{\partial u}{\partial y} = \frac{\partial^2 \phi}{\partial x \partial y} \quad \text{and} \quad \frac{\partial v}{\partial x} = \frac{\partial^2 \phi}{\partial x \partial y}$$

The existence of a velocity potential therefore implies irrotational flow.

(*b*) Since

$$u = \frac{\partial \psi}{\partial y}, \qquad \frac{\partial u}{\partial x} = \frac{\partial^2 \psi}{\partial x \partial y} \qquad \text{and}$$

since $$v = -\frac{\partial\psi}{\partial x}, \quad \frac{\partial v}{\partial y} = -\frac{\partial^2\psi}{\partial x \partial y}$$

then $$\frac{\partial u}{\partial x} + \frac{\partial v}{\partial y} = 0$$

This is one form of the continuity equation for steady incompressible flow in two dimensions.

Moreover, $$\frac{\partial u}{\partial x} = \frac{\partial^2\phi}{\partial x^2} \quad \text{and} \quad \frac{\partial v}{\partial y} = \frac{\partial^2\phi}{\partial y^2}$$

$$\therefore \qquad \frac{\partial^2\phi}{\partial x^2} + \frac{\partial^2\phi}{\partial y^2} = 0$$

This equation (well-known to mathematicians as the *Laplace equation*) and its derivation embraces the condition of continuity of incompressible fluids and the condition for irrotational flow. Any function which satisfies the Laplace equation is, therefore, a possible solution of an irrotational flow problem and the choice of function in any particular case is dictated by the boundary conditions.

It is left to the reader to satisfy himself that ψ also satisfies the Laplace equation, *i.e.* $\dfrac{\partial^2\psi}{\partial x^2} + \dfrac{\partial^2\psi}{\partial y^2} = 0$

(*c*) If $\phi = constant$, $\dfrac{\partial\phi}{\partial x} = 0$ and $\dfrac{\partial\phi}{\partial y} = 0$ and, therefore, the resulting velocity = 0. Hence $\phi = constant$ is an equipotential.

Similarly, it can be argued that $\psi = constant$ represents a streamline.

Three-dimensional flow

In general it is not possible to handle the mathematics of three-dimensional flow except for the special case of axi-symmetrical flow, which means that the flow is symmetrical about a single axis, say the x-axis.

EXAMPLE 33

(*a*) The velocity potential ϕ is defined by:

$$d\phi = -u\,dx - v\,dy$$

and the stream function ψ by:

$$d\psi = -v\,dx + u\,dy$$

Show that seepage flow through a porous medium obeying Darcy's law is a potential flow with $\phi = kH$ and $\psi = q$, where k is the permeability, H the total head, and q the two-dimensional flow rate.

(*b*) (*i*) Express u, v and ϕ in terms of x and y; and
(*ii*) sketch the flow net defined by $\psi = axy$;
(*iii*) quote an example of such a flow field in engineering practice.

SOLUTION

(*a*) Note that the sign convention for ϕ in this example is the opposite to that used in the text above.

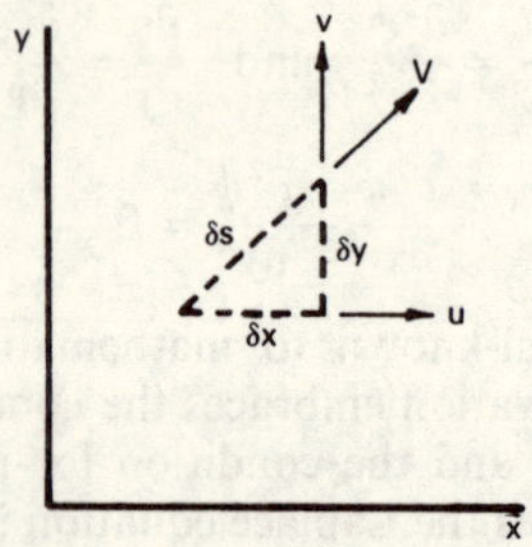

FIG. 53. Relationships between u and v and the resulting velocity V

With reference to Fig. 53, since:

$$-\mathrm{d}\phi = u\,\mathrm{d}x + v\,\mathrm{d}y$$

$$-\mathrm{d}\phi = V\mathrm{d}s$$

where V is the actual velocity at a point on the streamline and s is the distance to the point measured along the streamline from a convenient reference point 0.

Then
$$\phi_P = -\int_0^P V\mathrm{d}s$$

gives the value of ϕ at a point P on the streamline relative to its value at 0.

Darcy's law may be written $q = kli$ where q is the two-dimensional flow rate in m^2/s crossing an area $l \times 1$ and i is the hydraulic gradient.

$$\therefore \qquad \phi_P = -\int_0^P \frac{q}{l}\mathrm{d}s = -\int_0^P ki\,\mathrm{d}s = -k\int_0^P i\,\mathrm{d}s$$

since k is a constant. Also, $-\int_0^P i\,\mathrm{d}s$ is the head available between 0 and P

$$\therefore \qquad \underline{\underline{\phi = kH}}$$

If the distance along the normal to the streamline is n

$$\psi = \int V\mathrm{d}n = \int \frac{q}{l}\mathrm{d}n = q\int \frac{\mathrm{d}n}{l}$$

because q may be treated as constant since it has no component in the direction of n. Also, $\int \frac{dn}{l} = 1$ because l is the sum of the elements dn.

$$\therefore \qquad \psi = q$$

(*b*) (*i*) Since $\psi = axy$, $\quad u = \frac{\partial\psi}{\partial y} = ax$

and $\quad v = -\frac{\partial\psi}{\partial x} = -ay$

$$\therefore \qquad d\phi = ax\,dx + ay\,dy$$

and $\quad \phi = -\int ax\,dx + \int ay\,dy = \frac{a}{2}(-x^2 + y^2) + constant$

(*ii*) The streamlines and equipotentials are sketched in Fig. 54. The lines $\psi = constant = axy$ are hyperbolae in all four quadrants with the x- and y-axes asymptotic to $\pm\psi = a$. The equipotentials are also hyperbolae and are situated in the quadrants formed by the lines $y = \pm x$.

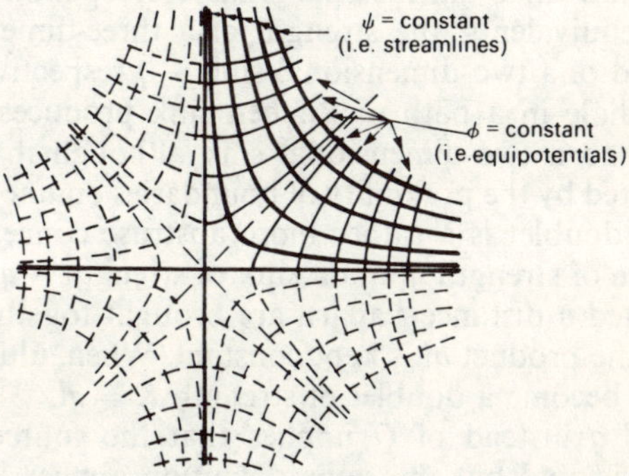

FIG. 54. *See* Example 33

(*iii*) With reference to the first quadrant only in Fig. 54, this type of flow will be found in practice in flow round a right-angle bend in an open channel—although it should be noted that the actual flow will depart from the theoretical near the boundaries because of the presence of a boundary layer.

STANDARD STREAM FUNCTIONS

A well-used method of determining flow patterns in theoretical hydrodynamics is by the use of certain standard stream functions. Four basic stream functions in two-dimensional flow exist for *uniform flow*, a *source*, a *sink* and a *doublet*, respectively; the last three being hypothetical. Descriptions of them are given below for both two-dimensional and three-dimensional cases, although the mathematical treatment is restricted to two dimensions.

Source. In three dimensions a source is a point from which flow emanates equally in all directions, the nearest thing to a practical example being the explosion of a bomb in otherwise still air.

A two-dimensional source is a point from which flow emanates equally in all directions *in one plane*. Sometimes, to distinguish between them, the three-dimensional case is called a *point source* and the two-dimensional case a *line source*.

The strength of a point source is defined as Q, the volumetric rate of flow from it, while the strength of a line source is defined as q (m^2/s), the volumetric rate of flow per unit length of the source. The streamlines for a line source are straight lines radiating outwards from the point representing the source and the equipotentials are concentric circles centred on the source.

Sink. A sink is a *negative* source, *i.e.* a point into which flow vanishes equally from all directions. Again, there are two-dimensional and three-dimensional sinks corresponding exactly to their source equivalents, the strength of a three-dimensional sink being $-Q$ and of a two-dimensional sink $-q$ respectively.

The drain hole in a bath or kitchen sink produces flow which approximates to a two-dimensional sink although it (the flow) is usually distorted by the proximity of boundaries on one side at least.

Doublet. A doublet is a rather more abstruse concept. It results when a source of strength q and a sink of strength $-q$, which were initially situated a distance l apart, are brought together in such a manner that the product ql is kept constant. When, ultimately, they coincide they become a doublet of strength $K = ql$.

The use of q instead of Q implies that the source (and sink) are two-dimensional but the same definition applies in the three-dimensional case.

1. ψ For uniform flow. Uniform flow conditions obtain when the velocity components u and v are the same at every point in the flow. Then:

$$\int d\psi = \int u\,dy - \int v\,dx$$

and

$$\underline{\underline{\psi = uy - vx}}$$

There is always an arbitrary constant of integration but this is not shown because ψ may also be considered to include an arbitrary constant.

2. *ψ For a source.* It is convenient to use polar co-ordinates r, θ, *see* Fig. 55, so that an element normal to the velocity vector is $rd\theta$ and,

$$\therefore \quad V = \frac{\partial \psi}{r \partial \theta}$$

$$\therefore \quad \psi = Vr\theta = \frac{q\theta}{2\pi}$$

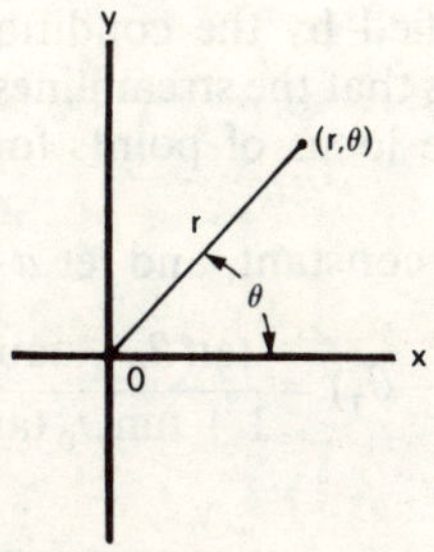

FIG. 55. Polar co-ordinates [*Note:* $r^2 = x^2 + y^2$ and $\tan\theta = y/x$, $x = r\cos\theta$ and $y = r\sin\theta$]

By convention, ψ is usually shown as ranging between 0 and $\pm\pi$.

3. *ψ For a sink.* A sink is a negative source and, therefore,

$$\psi = -\frac{q\theta}{2\pi}$$

Note: The relationships between rectangular co-ordinates and polar co-ordinates with the same origin are either:

$$r^2 = x^2 + y^2 \quad \text{and} \quad \tan\theta = \frac{y}{x}, \qquad \text{or}$$

$$x = r\cos\theta \quad \text{and} \quad y = r\sin\theta$$

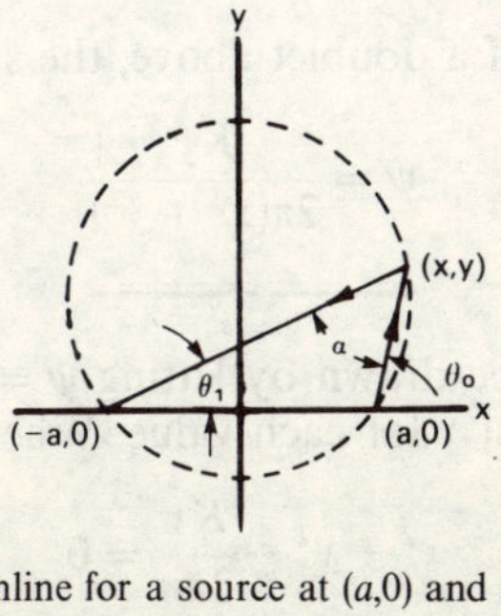

FIG. 56. Streamline for a source at $(a,0)$ and a sink at $(-a,0)$

4. *ψ For a doublet.*

Consider a source, strength q, situated at the point $(a,0)$ and a sink, strength $-q$, at $(-a,0)$ (*see* Fig. 56).

The stream function at any point (x,y) is obtained by adding algebraically the stream functions of the source and the sink, *i.e.*

$$\psi = \frac{q\theta_0}{2\pi} - \frac{q\theta_1}{2\pi} = \frac{q}{2\pi}(\theta_0 - \theta_1) = \frac{q\alpha}{2\pi}$$

Streamlines are identified by the condition that $\psi =$ constant, *i.e.* $\alpha =$ constant. It follows that the streamlines are circles, shown dotted in Fig. 56, because the locus of points for which α=constant is a circle.

Now let $2aq = K$, a constant, and let $a \to 0$. Then

$$\tan\alpha = \tan(\theta_0 - \theta_1) = \frac{\tan\theta_0 - \tan\theta_1}{1 + \tan\theta_0 \tan\theta_1}$$

$$= \frac{\dfrac{y}{x-a} - \dfrac{y}{x+a}}{1 + \dfrac{y^2}{x^2 - a^2}} = \frac{2ay}{x^2 + y^2 - a^2}$$

Since (x,y) is any point in the flow field the values of x and y are independent of a and, therefore, as $a \to 0$:

$$\frac{2ay}{x^2 + y^2 - a^2} \to 0 \quad \text{and} \quad \alpha \to \tan\alpha$$

Hence, as $a \to 0$, $$\alpha \to \frac{2ay}{x^2 + y^2}$$

and $$\psi = \frac{q\alpha}{2\pi} = \frac{q}{2\pi} \cdot \frac{2ay}{x^2 + y^2}$$

From the definition of a doublet above, the strength $K = 2aq$

$$\therefore \quad \psi = \frac{Ky}{2\pi(x^2 + y^2)}$$

The streamlines can be drawn by letting $\psi = 1, 2, 3, \ldots n, \ldots$ and then plotting x against y for each value. When $\psi = n$:

$$x^2 + y^2 - \frac{Ky}{2\pi n} = 0$$

which is the equation of a circle which touches the x-axis and has its centre at the point $x = 0$, $y = \dfrac{K}{4\pi n}$.

When $x = 0$,
$$y\left(y - \frac{K}{2\pi n}\right) = 0$$

and therefore, the ends of the diameter are at $y = 0$ and $y = \dfrac{K}{2\pi n}$.
The streamlines are as indicated in Fig. 57.

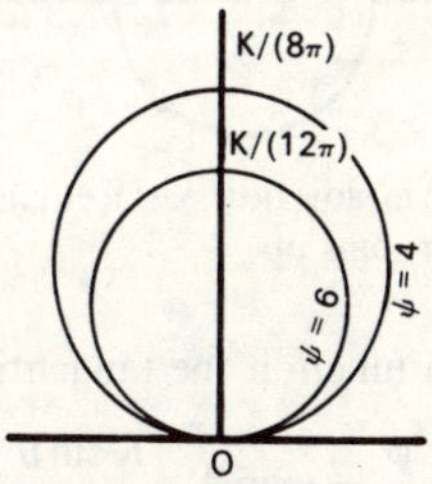

FIG. 57. Streamlines for a doublet

EXAMPLE 34

(*a*) From the stream functions describing (*i*) a uniform flow; (*ii*) a source; (*iii*) a sink; and (*iv*) a doublet, derive the equation of the flow around a circular cylinder.

(*b*) Deduce the equation of the tangential velocity on the cylinder surface.

SOLUTION

(*a*) The required stream functions are given immediately above. In a theoretical treatment such as this, the direction of the uniform flow is arbitrary and for convenience, it is chosen here as being parallel to the x-axis with flow from right to left. For this flow, $\psi = -uy$ (with $v = 0$). Using the stream function derived above for a doublet, the combined flow is:

$$\psi = -uy + \frac{Ky}{2\pi(x^2 + y^2)}$$

Using polar co-ordinates, $y = r\sin\theta$, $x^2 + y^2 = r^2$ and $\dfrac{y}{x^2 + y^2} = \dfrac{\sin\theta}{r}$

$$\therefore \quad \psi = -ur\sin\theta + \frac{K\sin\theta}{2\pi r}$$

$$\psi = 0 \quad \text{when} \quad 2\pi r^2 u = K$$

i.e. when
$$r = \sqrt{\frac{K}{2\pi u}}$$

Thus one of the streamlines, $\psi = 0$, is a circle with its centre at the origin and with radius $= \sqrt{\dfrac{K}{2\pi u}}$. Elsewhere, $\psi = 0$ is the x-axis and, therefore, the streamline $\psi = 0$ is as shown in Fig. 58. Hence, the stream function $\psi = -ur\sin\theta + \dfrac{K\sin\theta}{2\pi r}$ represents a uniform flow past a circular cylinder.

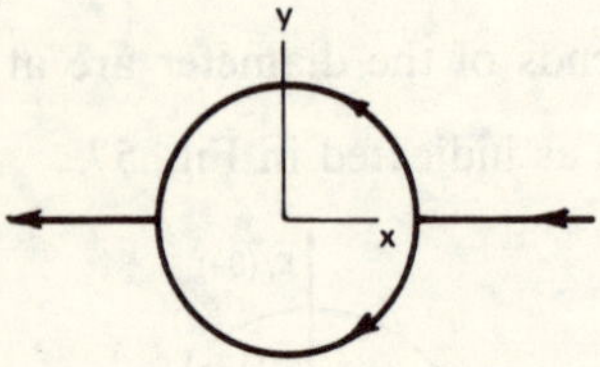

FIG. 58. The dividing streamline in slow flow past a circular cylinder. *Stagnation points* occur where the flow divides or joins up

(*b*) For the above stream function the tangential velocity is:

$$-\frac{\partial\psi}{\partial r} = u\sin\theta + \frac{K\sin\theta}{2\pi r^2}$$

On the surface, $r^2 = \dfrac{K}{2\pi u}$

$\therefore$ tangential velocity $= \underline{\underline{2u\sin\theta.}}$

NOTE ON EXAMPLE 34

By rearranging the stream function formula for the given flow thus:

$$2\pi ur^2\sin\theta + 2\pi r\psi - K\sin\theta = 0$$

and solving this as a quadratic in r, it will be found that

$$r = \frac{-\psi \pm \sqrt{\psi^2 + 2\sin^2\theta\, Ku/\pi}}{2u\sin\theta}$$

$$= \frac{-\psi + \sqrt{\psi^2 + 2\sin^2\theta\, Ku/\pi}}{2u\sin\theta}$$

when the negative root, which is meaningless, is ignored.

From this expression for r, it follows that:

if ψ is negative and large, r is large and the streamlines will lie *outside* the cylinder and thus represent flow outside the circle of radius $\sqrt{K/(2\pi u)}$;

if ψ is positive, $r \to 0$ as ψ increases and, therefore, positive values of ψ give streamlines *inside* the circle of radius $\sqrt{K/(2\pi u)}$.

Hence, by ignoring positive values of ψ, the stream function $\psi = -uy + Ky/\{2\pi(x^2 + y^2)\}$ represents flow past the circular cylinder of radius $\sqrt{K/(2\pi u)}$ and is the equation from which the streamlines can be plotted.

EXAMPLE 35

Using the results of Example 34 plot streamlines to represent flow past a circular cylinder. Assume a uniform flow, remote from the cylinder, of 4 m/s in the direction of x decreasing and a doublet of strength 8π.

SOLUTION

Here, $u = 4$ and $K = 8$ and, therefore, the radius of the cylinder is:

$$r = \sqrt{\frac{8\pi}{2\pi \times 4}} = 1$$

The dividing streamline comprises this circle and the x-axis outside it. The other streamlines could be plotted, laboriously, direct from the equation for ψ, namely, $\psi = -4y + \frac{8\pi y}{2\pi(x^2 + y^2)}$, by ascribing negative values to ψ and, for each, plotting x *vs* y.

A simpler method, however, is to plot the two contributing flows independently on the same co-ordinate axes and then join the intersections which have the same combined ψ-values. The steps are as follows.

(*i*) Choose a suitable point as origin, near the centre of the graph paper, and mark off on the y-axis distances which represent intervals of one metre. A suitable scale is 2 *cm* to represent 1 *metre.*

(*ii*) Make another vertical scale of the values of ψ_1, the stream function for the uniform flow, from $\psi_1 = -4y$ and draw feint horizontal lines for $\psi_1 = 0, -1, -2, -3,$ *etc.*

(*iii*) Calling the stream function for the doublet ψ_2 and remembering that $K = 8\pi$,

$$\psi_2 = \frac{4y}{x^2 + y^2}$$

i.e.

$$\psi_2(x^2 + y^2) - 4y = 0$$

which, when $x = 0$ reduces to

$$y(\psi_2 y - 4) = 0$$

i.e.

$$y = 4/\psi_2 \quad \text{or} \quad y = 0$$

The streamlines for a doublet, remember, are circles which touch the x-axis and which have their diameters lying along the y-axis. The above values of y give the extremities of the diameter. Hence a table is prepared below from which the circles can be plotted.

ψ_2 :	4·0	3·0	2·0	1·8	1·6	1·4	1·2	1·0	0·5
diameter:	1·00	1·33	2·00	2·22	2·50	2·86	3·33	4·00	8·00

At each intersection of the two flow patterns the values of ψ_1 and ψ_2 are added to give ψ at the intersection. Streamlines for flow past the cylinder can then be obtained by joining with a smooth curve the intersections which have the same ψ-values, the result being shown in Fig. 59.

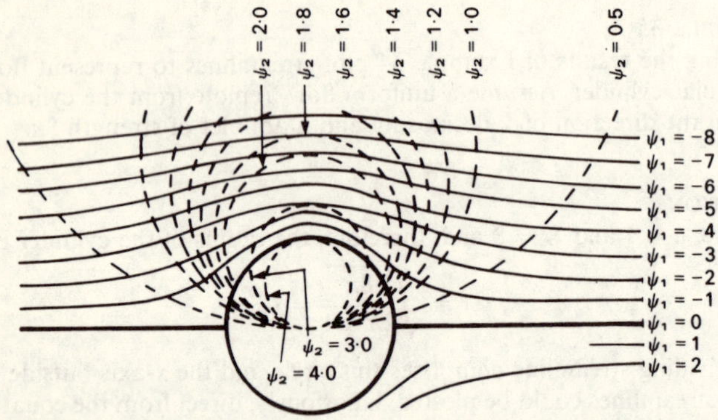

FIG. 59. Flow past a circular cylinder. *See* Example 52. Lines for $\psi_1 = -1, -2, -3, 7, 8$ not shown

VORTICES AND CIRCULATION

Although other types of vortex can be identified in fluid mechanics, the most fundamental are the forced vortex and free vortex. *Circulation* is a flow property, relevant to both, which is closely connected with *lift*, a phenomenon dealt with further in Chapter 11, p. 175. It also has a direct relationship with *rotation* an introduction to which may help (by contrast) towards a fuller understanding of irrotational flow.

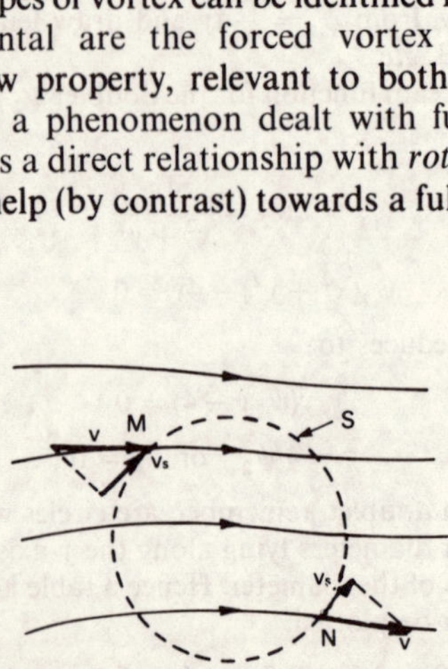

FIG. 60. Illustrating the definition of circulation

Circulation

Figure 60 shows a closed path S superimposed on a pattern of streamlines with M and N marking two typical intersections. At each of these a velocity triangle is drawn showing the velocity v on the streamline and its component v_s tangential to S. It will be noticed that at M, v_s is clockwise and at N, v_s is counter-clockwise, although this need not necessarily always be the case. If δs, an, element of S, is envisaged at all points such as M and N then circulation can be defined as the sum of the products $v_s \times \delta s$ taken round the whole of S. Denoting circulation by Γ (gamma) this can be written mathematically thus:

$$\Gamma = \int_S v_s \mathrm{d}s \quad \text{or} \quad \Gamma = \oint v_s \mathrm{d}s$$

where $\int_S$ and $\oint$ are alternative ways of signifying *integral over the closed surface* S. If, for example, a liquid element moves in a circle of radius r with an angular velocity of ω (omega), then at every point of the circle $v_s = \omega r$ and $\Gamma = \int_S \omega r \mathrm{d}s = \int_0^{2\pi} \omega r . r\mathrm{d}\theta$.

i.e. $= 2\pi\omega r^2$

an important result which will be referred to again.

Except in certain special cases the existence of circulation implies that the flow is rotational and, in fact, ω is the *rotation*. In rotational flow, moreover, each fluid element rotates about an axis within itself. $\Gamma = 0$ signifies *irrotational* flow.

Another property directly related to circulation is vorticity ξ (chi) which is defined as the *circulation per unit area*. In the above example the area enclosed by the circle is πr^2 and, therefore:

$$\xi = \frac{\Gamma}{\pi r^2} = 2\omega$$

a result which is true for all rotational flows. Strictly, a more accurate definition of vorticity is $\xi = \underset{A \to 0}{\text{Limit}}\left(\frac{\Gamma}{A}\right)$ where A is the cross-sectional area of a fluid element with rotation ω.

Vortices

Before dealing separately with the two types of vortex there are some preliminary results to be established which are common to both.

Preliminaries. In both vortices, the fluid elements move in circles about the centre of the vortex. If the angular velocity of the motion is ω at radius r, the tangential velocity is (now dropping the subscript s) $v = \omega r$.

$$\therefore \qquad \Gamma = \int_0^{2\pi} \omega r, \, n \mathrm{d}\theta = 2\pi\omega r^2$$

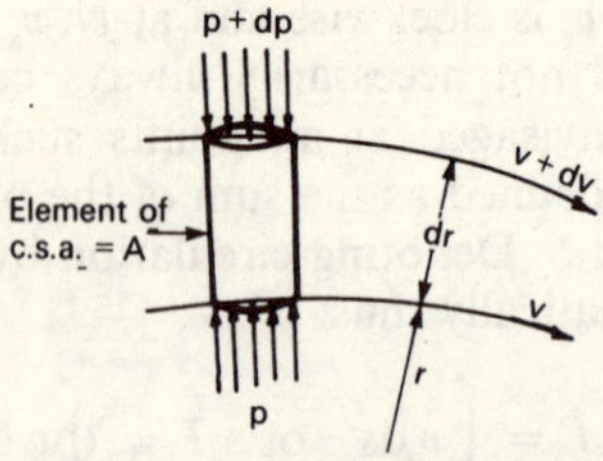

FIG. 61. Showing the pressures acting on an element of fluid when it is moving in a curved path

Figure 61 illustrates the forces acting on an element of fluid which is moving in a curved path bounded by two streamlines lying in the same horizontal plane. The flow velocities at the two streamlines are v and $v + \mathrm{d}v$ respectively and the distance between them is $\mathrm{d}r$. Then:

mass of element $= \rho A \, \mathrm{d}r$;

net horizontal force acting on it $= A \, \mathrm{d}p$ inwards; and

inward radial acceleration $= \dfrac{v^2}{r}$

$$\therefore \qquad A \, \mathrm{d}p = \frac{v^2}{r} \times \rho A \, \mathrm{d}r$$

$$\therefore \qquad \frac{\mathrm{d}p}{\mathrm{d}r} = \rho \frac{v^2}{r}$$

Applying the Bernoulli equation along a horizontal streamline:

$$H = \frac{p}{\rho g} + \frac{v^2}{2g}$$

$$\frac{\mathrm{d}H}{\mathrm{d}r} = \frac{1}{\rho g} \times \frac{\mathrm{d}p}{\mathrm{d}r} + \frac{2v}{2g} \times \frac{\mathrm{d}v}{\mathrm{d}r}$$

$$= \frac{1}{\rho g} \times \frac{\rho v^2}{r} + \frac{v}{g} \times \frac{\mathrm{d}v}{\mathrm{d}r}$$

$$\underline{\frac{\mathrm{d}H}{\mathrm{d}r} = \frac{v}{g}\left(\frac{v}{r} + \frac{\mathrm{d}v}{\mathrm{d}r}\right)}$$

Forced vortex

If a vertical cylinder containing liquid is rotated about its axis, viscous shear will develop from the boundary inwards causing the fluid to rotate too. If a steady rotation is kept up long enough all the contained liquid will be moving like a solid mass with the same angular velocity (ω say) as the cylinder. Every fluid element is then turning about the axis of rotation and, consequently, turns about an axis within itself with the same angular velocity ω. This type of flow is called a *forced* vortex or, sometimes, a *solid body* vortex and is an example of rotational flow. Another practical example of a forced vortex is to be found in the action of a paddle-steamers paddle. The water in direct contact with the paddle behaving like a *forced* vortex although the water outside the direct influence of the paddle tends to behave more in the nature of a *free* vortex as the rotation is attenuated by viscous shear.

The salient feature of a forced vortex is that every fluid element rotates; if the rotation ω is constant, the tangential velocity v at radius r is $v = \omega r$:

$$\frac{v}{r} = \omega \quad \text{and} \quad \frac{dv}{dr} = \omega$$

$$\frac{dH}{dr} = \frac{\omega r}{g}(\omega + \omega) = \frac{2\omega^2 r}{g}$$

Integrating:

$$H = \frac{\omega^2 r^2}{g} + C$$

in which the constant of integration C can be made zero by choosing the datum so that $H = 0$ when $r = 0$.

Free vortex

A free vortex is a mathematical concept of an irrotational flow for which the streamlines are all circles centred on the same vertical axis. Because the flow is irrotational a velocity potential must exist. In polar co-ordinates the tangential velocity is $\frac{1}{r}\frac{\partial \phi}{\partial \theta}$ and the velocity normal to the streamline is $\frac{\partial \phi}{\partial r} = 0$. The velocity potential ϕ is a function of r, θ which, therefore, must satisfy simultaneously the equations:

$$\frac{\partial \phi}{\partial r} = 0 \quad \text{and} \quad \frac{1}{r}\frac{\partial \phi}{\partial \theta} = \omega r$$

Since $\frac{1}{r}\frac{\partial \phi}{\partial \theta} = \omega r$, $\phi = \omega r^2 \theta + f(r)$.

Since $\dfrac{\partial \phi}{\partial r} = 0$, ϕ must be independant of r and, therefore, $f(r) = 0$ (or a constant), $\omega r^2 = \mu$, where μ is a constant called the *strength* of the vortex.

Since $v = \omega r$, $$v = \frac{\mu}{r}$$

$$\frac{v}{r} = \frac{\mu}{r^2} \text{ and } \frac{dv}{dr} = -\frac{\mu}{r^2}$$

$$\therefore \quad \frac{dH}{dr} = \frac{v}{g}\left(\frac{v}{r} + \frac{dv}{dr}\right) = 0$$

Integrating,

$$H = \text{constant}$$

EXAMPLE 36

(*a*) Derive expressions for the free surface of a forced vortex and a free vortex.

(*b*) Distinguish between rotational and irrotational flow and explain the significance of velocity potential in this connection. Does a velocity potential exist for both a forced vortex and a free vortex?

(*c*) Explain what is meant by circulation and its relevance to irrotational flow, making a particular reference to a free vortex.

SOLUTION

(*a*) For a forced vortex, $H = \dfrac{\omega^2 r^2}{g}$ and ω is a constant. On a streamline, applying the Bernoulli equation:

$$H = \frac{p}{\rho g} + \frac{v^2}{2g} + z$$

At the free surface, $p = 0$.

$$\therefore \quad \frac{\omega^2 r^2}{g} = \frac{v^2}{2g} + z = \frac{\omega^2 r^2}{2g} + z$$

$$\therefore \quad z = \frac{\omega^2 r^2}{2g} = \text{constant} \times r^2$$

which is the equation of the free surface, *see* Fig. 62 (*a*).

For a free vortex, $H = \dfrac{p}{\rho g} + \dfrac{v^2}{2g} + z = \text{constant}$.

At the free surface $p = 0$ and, therefore, $H - z = \dfrac{\omega^2 r^2}{2g}$.

Also, for a free vortex, $\omega = \dfrac{\mu}{r^2}$

$$H - z = \frac{\mu^2}{2gr^2} = \frac{\text{constant}}{r^2}$$

which is the equation of the free surface, *see* Fig. 62 (*b*).

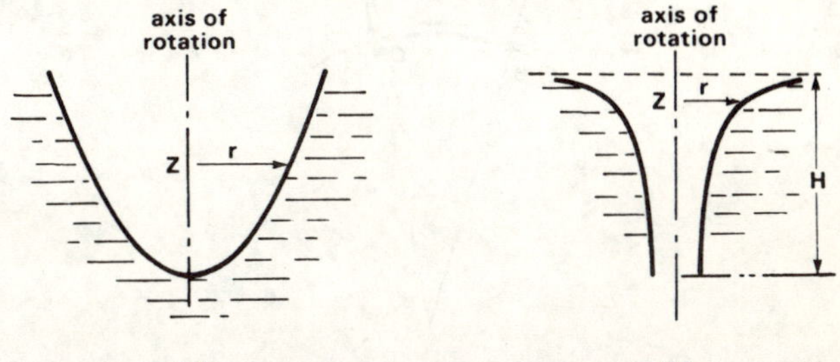

FIG. 62. Sections through the free surfaces of (*a*) forced vortex and (*b*) free vortex

(*b*) The description of how rotational flow differs from irrotational is given on p. 123. In addition the two other important differences should be stated, namely:

(*i*) For irrotational flow a velocity potential ϕ always exists whereas for rotational flow there can be no velocity potential.

For both a free vortex and forced vortex a velocity potential, if it exists, must satisfy simultaneously the same conditions, $\dfrac{1}{r}\dfrac{\partial\phi}{\partial\theta} = \omega r$ and $\dfrac{\partial\phi}{\partial r} = 0$. The only solution is $\phi = \omega r^2\theta$ which satisfies the conditions for a free vortex because, in its case, $\omega r^2 = \mu$ and, therefore, $\phi = \mu\theta$. Clearly there is no such solution for a forced vortex which is characterised by $\omega =$ constant.

(*ii*) They also differ in that if circulation exists, the flow is rotational whereas it cannot exist in the case of irrotational flow.

(*c*) Circulation is defined on p. 123. In irrotational flow, $\Gamma = 0$. There is an apparent paradox in the case of a free vortex, for which $\Gamma = 2\pi\omega r^2 = 2\pi\mu$, because a free vortex is also irrotational flow. The explanation is that a free vortex is irrotational at every point other than its centre where the rotation is ω and, in fact, if Γ is calculated for any closed path in a free vortex, which does not include the centre, the circulation would be found to be zero, *see* below. The centre of the free vortex is called a *singular point* because when $r = 0$, $v = \dfrac{\mu}{r} = \infty$.

With reference to Fig. 63, the circulation round *ABCD*, for both types of vortex, is

$$\Gamma = \int_0^{\alpha} (\omega_2 r_2{}^2 - \omega_1 r_1{}^2)\mathrm{d}\theta = (\omega_2 r_2{}^2 - \omega_1 r_1{}^2)\alpha$$

For a free vortex, $\omega_2 r_2^{\,2} = \omega r_1^{\,2} = \mu$ and, therefore, $\Gamma = 0$.

For a forced vortex, $\omega_2 = \omega_1 = \omega$, say, and therefore, $\Gamma = \omega(r_2^{\,2} - r_1^{\,2})\alpha$.

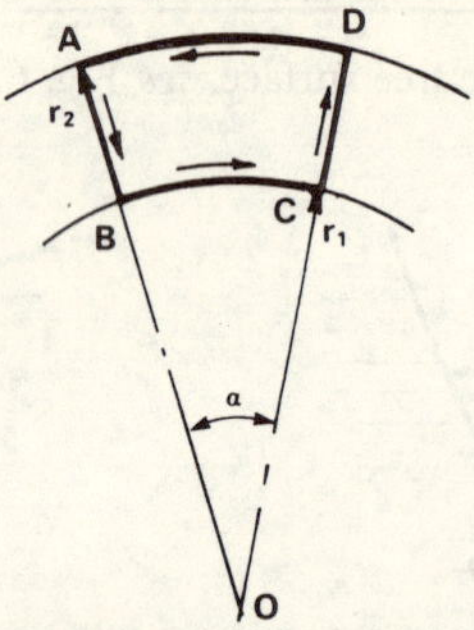

FIG. 63. Showing a circulation round *ABCD* which does not include *O*

SUMMARY

In Table II a summary is given of the standard stream functions and velocity potentials and other pertinent formulae. A good exercise for the reader is to validate the expressions given for ψ and ϕ, not all of which appear in the text, by satisfying himself that they yield velocity components which satisfy the appropriate flow pattern.

Flow pattern	*Description*	ψ	ϕ
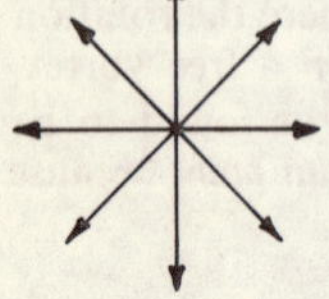	Uniform flow	$\psi = uy - vx$	$\phi = ux + vy$
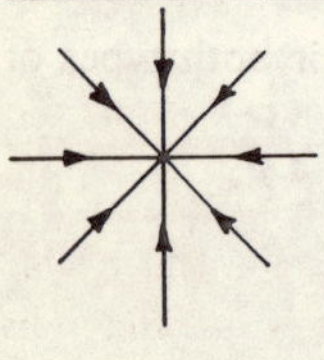	Source, strength q	$\psi = \dfrac{q\theta}{2\pi}$	$\phi = \dfrac{q}{2\pi}\ln r$
	Sink, strength $-q$	$\psi = -\dfrac{q\theta}{2\pi}$	$\phi = -\dfrac{q}{2\pi}\ln r$

Flow pattern	*Description*	ψ	ϕ
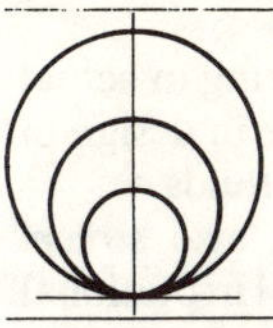	Doublet, strength K	$\psi = \dfrac{Ky}{2\pi(x^2 + y^2)}$	$\phi = \dfrac{-Kx}{2\pi(x^2 + y^2)}$
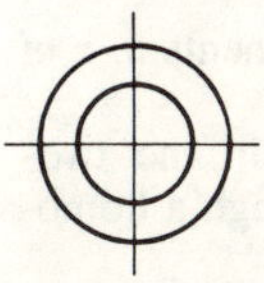	Free vortex	$\psi = -\frac{1}{2}\omega r^2$	$\phi = \mu\theta$
	Forced vortex	$\psi = -\frac{1}{2}\omega r^2$	——

TABLE II: Summary of standard cases of ψ and ϕ in two-dimensional flow

Notes:

Convention used: ψ: flow is from left to right when looking in the direction of ψ increasing

ϕ: increases in direction of flow.

Cartestian co-ordinates: $u = \dfrac{\partial\psi}{\partial y} = \dfrac{\partial\phi}{\partial x}$

$$v = -\frac{\partial\psi}{\partial x} = \frac{\partial\phi}{\partial y}$$

Polar co-ordinates: radial velocity $= \dfrac{\partial\phi}{\partial r} = \dfrac{\partial\psi}{r\partial\theta}$

$$\text{tangential velocity} = \frac{\partial\phi}{r\partial\theta} = -\frac{\partial\psi}{\partial r}$$

QUESTIONS

1. Plot, in the manner described on p. 121 the streamlines for the combined flow in two dimensions given by a source of strength 16 m^2/s and a uniform stream with a velocity $u = 1$ m/s.

A convenient scale for the uniform flow streamlines is 10 mm to an increment of 2 in the ψ-values.

2. (*a*) Define *one-dimensional* and *two-dimensional* flow. For a two-dimensional flow field define the velocity potential ϕ and the stream function ψ and relate them to the lines forming the meshes of a flow net.

(*b*) Ground water seeps at a discharge velocity of 1 m/day through a homogeneous isotropic confined aquifer of 10 m thickness. A vertical well is sunk to the base of the aquifer. Sketch the flow net and calculate the equation of the envelope of the flow drawn to the well if water is pumped out at a steady rate of 1000 m^3/day.

3. (*a*) Define *ideal* fluids. Also define *circulation* and *vorticity* in a two-dimensional flow field.

(*b*) In a two-dimensional flow, $u = 2z$ and $w = 2x$ (where z is the Cartesian co-ordinate dimension measured vertically upwards and u and w are the velocity components in the x and z directions respectively). Sketch the flow net, calculate the general expressions for the potential function ϕ and the stream function ψ, and show that the flow is irrotational.

(*c*) Give an example of a part of the flow net in (*b*) occurring in actual engineering practice and state how allowance is made in design in such a case for the difference between ideal and real fluids.

4. (*a*) Define *streamlines*, *equipotential lines* and *flow nets*; also *stream functions* ψ and *potential functions* ϕ in two-dimensional irrotational flow.

(*b*) State the relations between ψ and ϕ and the components u, v of the total velocity V at any point x, y.

(*c*) What physical variables correspond to ψ and ϕ in the two-dimensional flow of a (Newtonian) viscous fluid through a homogeneous isotropic porous medium?

(*d*) Oil is pumped at a steady rate Q from an extensive horizontal reservoir rock of permeability k and thickness D, sandwiched between shale strata, through a bore hole of diameter d in which the free oil surface is above the top of the reservoir rock, *see* Fig. 64. Derive the equations of ψ and ϕ, or of their physical equivalents, in terms of k, D, Q and of the polar co-ordinates r, θ centred on the bore hole.

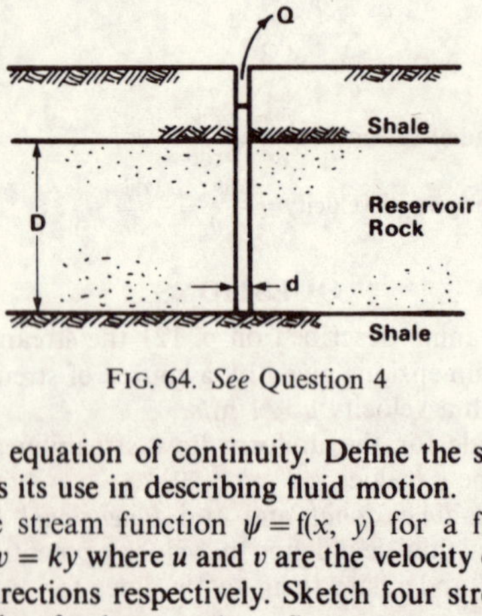

FIG. 64. *See* Question 4

5. (*a*) Derive the equation of continuity. Define the stream function ψ, and discuss its use in describing fluid motion.

(*b*) Derive the stream function $\psi = f(x, y)$ for a flow described by $u = -kx$, $v = ky$ where u and v are the velocity components in the x and y directions respectively. Sketch four streamlines and give an example of where such a flow pattern could exist in any typical system.

6. (*a*) Define *vorticity* and discuss the significance of irrotational motion.

(*b*) Give the vorticity at 1 m and 3 m radius in a vortex whose speed is 1 m/s throughout, and calculate the difference in pressure between these two places if the axis of rotation is (*i*) vertical and (*ii*) horizontal.

7. (*a*) Define a *free vortex*.

(*b*) The inside and outside walls of a rectangular channel describe concentric circular arcs of 5 m and 7 m radii respectively. The velocity distribution in the bend is approximated by a free vortex, with the maximum velocity on the inside wall equal to 3 m/s. Calculate the superelevation in the water surface.

8. (*a*) Define *one-dimensional* and *two-dimensional* flow. For a two-dimensional flow define the velocity potential ϕ and the stream function ψ and relate them to energy potential and two-dimensional flow rate in a potential flow.
   (*b*) Show that laminar seepage flow of water through an aquifer represents potential flow.
   (*c*) The steady discharge per unit width of a horizontal artesian aquifer of thickness D is q. A discharge Q flows out of a borehole of diameter d penetrating to the base of the aquifer. Derive the equations of the equipotential lines and of the stream lines of the artesian flow in any horizontal plane.

9. (*a*) Define the stream function ψ, the potential function ϕ, and the vorticity ξ. What is the physical meaning of the statement $\psi(x, y, z) = constant$?
   (*b*) The Cartesian components of a total velocity V are u, v, w in the x, y, z directions. Derive the equations of the streamlines of the following patterns:

   (*i*) $u = 20y, \quad v = 0 \quad w = 1;$
   (*ii*) $u = y, \quad v = x \quad w = 0.$

   Sketch characteristic streamlines of each, check whether the flows are irrotational and indicate if such flows could occur in nature or in some engineering device.

10. Figure 65 shows a steady, uniform, two-dimensional flow described by the stream function $\psi = uy - vx$.

    (*a*) Show that such a flow can exist.
    (*b*) Prove that the circulation round the rectangular circuit *ABCD* is zero.
    (*c*) Derive the equation describing the velocity potential of the flow and sketch typical equipotential lines.

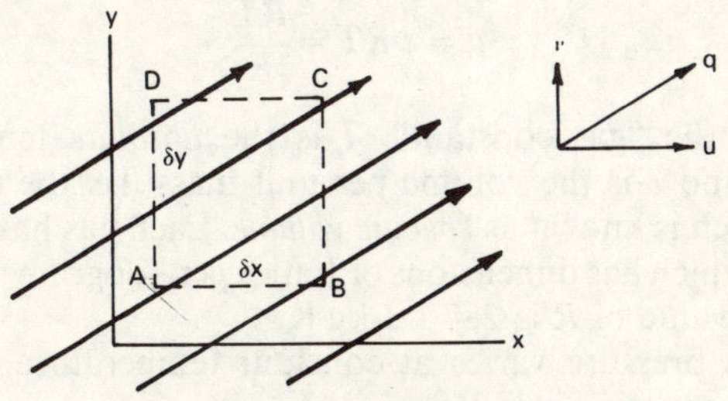

FIG. 65. *See* Question 10

CHAPTER 9

COMPRESSIBLE FLOW

MOST of the problems which have been considered up to this point have assumed that the fluid is of constant density. This assumption greatly simplifies the equations which are used and is valid for many practical cases. In reality all fluids are compressible to some extent, and one particular case of *liquid* compressibility (water hammer) has already been encountered. However, most of the compressible flow problems in engineering occur when the fluid is a gas. This is because the density of a gas is far more sensitive to changes of pressure and temperature than the density of a liquid. This chapter, therefore, concentrates on gas flow processes in which there are substantial changes of fluid density. Typical cases of such flows occur in air compressors and gas turbines, and also with gas flows in pipelines.

In order to analyse these processes it is necessary to employ the principles of thermodynamics, of which a brief outline is given below. For a fuller discussion of these concepts, and for tables of the thermodynamic properties of fluids, a comprehensive textbook on thermodynamics should be consulted.

THERMODYNAMIC CONCEPTS

1. The basic equation of state of a "perfect" gas is

$$p = \rho RT = \frac{RT}{v} \tag{1}$$

where R is the "gas constant", T is the *absolute* temperature in degrees K, and v is the volume per unit mass (i.e. the reciprocal of density) which is known as *specific volume*. Each gas has a particular value of R, which has dimensions of Joules per kilogram per degree K. For air the value of R is 287·1 J/kg K.

If the gas pressure varies at constant temperature equation (1) becomes an expression of Boyle's law:

$$\frac{p}{\rho} = pv = \text{constant}$$

Similarly if the temperature of a gas is varied whilst pressure is held constant, equation (1) reduces to Charles' law:

$$\frac{1}{\rho T} = \frac{v}{T} = \text{constant}$$

Whilst real gases do not exactly follow "perfect" gas laws, calculations based on these relationships are usually sufficiently accurate for most engineering purposes.

2. *Heat* is a form of energy which can only be transferred from one system to a second system when there is a difference in their temperatures. If the second system is at the lower temperature, then it would receive a *positive* heat transfer whilst there would be *negative* heat transfer from the first system. The amount of heat energy transferred is measured in Joules and the symbol used is Q. (Note that the same symbol is used for both heat transfer *and* for rate of discharge of liquids.) Alternatively, heat transferred to unit mass of gas ($= Q/m$) has the symbol q.

Each substance possesses a *specific heat* which is defined as the amount of heat supplied to unit mass of the substance in order to raise its temperature through one degree. For gases not one but two such values are defined. Consider the system shown in Fig. 66, which shows a volume of gas contained in a cylinder with a piston.

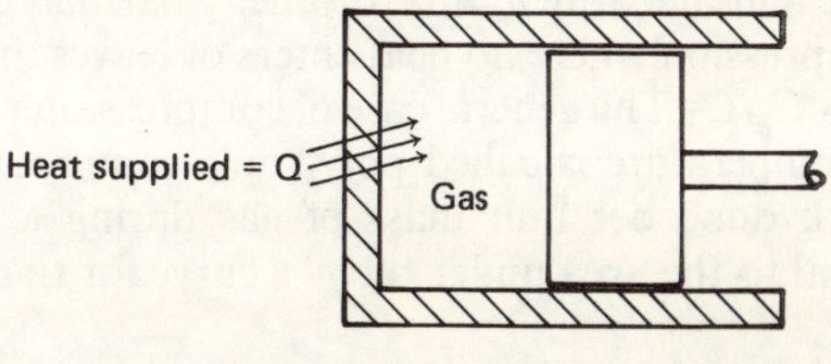

FIG. 66

(*a*) If heat Q is transferred to the gas while the volume is held constant, then no work is done by the gas, since the performance of work would require that the piston should move. All the heat transferred must be employed in increasing the energy of the gas itself, and this process is known as an increase in the *internal energy* (e) of the gas. The specific heat of a gas at constant volume has the symbol C_v and is defined by

$$C_v = \frac{\mathrm{d}e}{\mathrm{d}T} \text{ (J/kg.K)}$$

Thus the amount of heat transferred to mass m of the gas is $\mathrm{d}Q = mC_v\mathrm{d}T$.

(*b*) If heat is transferred to the gas, and the piston is allowed to move in such a way that the pressure is held constant, then work is performed. The heat energy transferred is then partly absorbed in work and partly in increasing the internal energy of the gas. The sum of internal energy plus work done is known as the *enthalpy h* of the gas. The specific heat of a gas at constant pressure has the symbol C_p and is defined as

$$C_p = \frac{dh}{dT} \text{ (J/kg.K)} \qquad \text{where } h = e + pv$$

The amount of heat transferred is $dQ = mC_p dT$. Clearly C_p must be greater than C_v.

(*c*) There are two important relationships between C_p and C_v: (i) The difference $(C_p - C_v) = R$. (ii) The ratio $C_p/C_v = \gamma$. As an example, the value of γ for air is 1·4, and the use of γ will be demonstrated in some of the examples.

3. The relationship between pressure p and specific volume v, when a gas is compressed, has been investigated extensively. This experimental work has shown that the relationship between p and v is approximately of the form pv^n = const. The value of n, which is known as the polytropic index, depends on the nature of the compression process. One particular example is isothermal (constant temperature) compression, for which it has already been shown that pv = constant, and therefore $n = 1$. Another particular case is that of adiabatic compression where no heat enters or leaves the gas and for which $n = \gamma = C_p/C_v$. The general case of compression with heat flow and varying temperature is called polytropic compression.

4. The work done per unit mass of gas during a compression process is equal to the area under the p–v curve for that process, i.e. $w = \int_1^2 p\, dv$.

5. The *first law of thermodynamics* states that for a thermodynamic cycle, net heat supplied to the system must equal the sum of the net external work done plus the change of internal energy of the working fluid. This may be expressed:

$$\sum dq = \sum dw + de$$

where $q = Q/m$ = heat supplied per unit mass of fluid, w = work done per unit mass of fluid, and $\sum$ is the summation for one "cycle" of a compression process. By a thermodynamic "cycle" we mean that the fluid passes through a series of processes and finally returns to its original state.

6. The *second law of thermodynamics* states that for a system to operate in a thermodynamic cycle and produce work it must exchange or transfer heat between "hot" and "cold" heat reservoirs. This is illustrated in Fig. 67.

A well-known example of such a process is that of the steam turbine

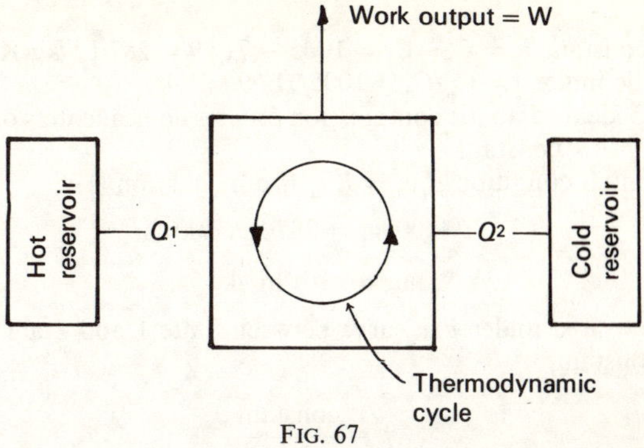

FIG. 67

system. Fuel is burnt in a boiler furnace (hot reservoir) the heat (Q_1) from which is used to transform water into high-pressure steam. The steam performs a thermodynamic cycle by (i) passing from the boiler to a turbine in which much of the energy in the steam is converted into a mechanical work output, (ii) passing from the turbine to a cooling condenser (cold reservoir) which extracts heat (Q_2) from the steam and converts the steam back to water, (iii) passing from the condenser back to the boiler.

7. The property of entropy (s) arises from the second law and is defined by the equation $T\mathrm{d}s = \mathrm{d}q$. The dimensions of entropy are J/kg.K.

Strictly speaking, this equation applies only to a *reversible* process, i.e. a process in which a gas is compressed and then decompressed in such a way that the gas, compressor and the surrounding environment all return to their original state. In practice, due to friction and other losses, no process is truly reversible. Nevertheless entropy is extremely useful since, being a *property*, it can be related to the other properties (p, T, v or ρ) of each gas and presented in diagrammatic or tabular form. For an ideal, reversible adiabatic process entropy remains constant and hence such a process is often known as an isentropic process. For any irreversible process entropy increases.

EXAMPLE 37

Using the values of the specific heats for air which are given below, calculate the gas content R and the index γ for adiabatic compression.

1 kg of air is compressed from its initial pressure and temperature of 100 kN/m^2 and 300 K to 500 kN/m^2. Compute the work done and heat supplied during compression if the compression process is (a) frictionless adiabatic, (b) polytropic with a compression index of 1·2.

$$C_p = 1005 \text{ J/kg.K} \qquad C_v = 717{\cdot}9 \text{ J/kg.K}$$

SOLUTION

The gas constant $R = C_p - C_v = 1005 - 717{\cdot}9 = 287{\cdot}1$ J/kg.K.
The adiabatic index $\gamma = C_p/C_v = 1005/717{\cdot}9 = 1.4$.

(*a*) For an ideal adiabatic compression process no heat enters or leaves the system and pv^γ = constant.

For the initial condition $p_1 v_1 = RT_1$ (from equation (1))

$$100\,000 v_1 = 287{\cdot}1 \times 300$$

$$\therefore\ v_1 = 1{\cdot}161\ \text{m}^3/\text{kg}$$

Work done = area under p–v curve between State 1 and State 2 = $\int_1^2 p\,dv$ but pv^γ = constant

$$\therefore\ p = \frac{\text{constant}}{v^\gamma}$$

As both p_1 and v_1 are known, the constant may be evaluated:

$$p_1 v_1{}^\gamma = 100\,000 \times 1{\cdot}161^{1{\cdot}4} = 123\,244$$

Also $p_2 v_2{}^\gamma = p_1 v_1{}^\gamma$, i.e. $500\,000 v_2{}^{1{\cdot}4} = 100\,000 \times 1{\cdot}161^{1{\cdot}4}$;
hence $v_2 = 0{\cdot}3678$ m³/kg

$$\therefore\ \text{work done} = \int_1^2 p\,dv = \int_1^2 \frac{123\,244}{v^{1{\cdot}4}}.dv$$

$$= \int_{1{\cdot}161}^{0{\cdot}3678} \frac{123\,244}{v^{1{\cdot}4}} = -169\,436\ \text{J/kg}$$

The negative sign indicates that work is done on the gas (whereas a positive sign would indicate work done *by* the gas).

(*b*) For the polytropic case pv_1 = const.

The initial volume is computed as in part (*a*) above, $\therefore v_1 = 1{\cdot}161$ m³/kg.

$$p_1 v_1{}^{1{\cdot}2} = p_2 v_2{}^{1{\cdot}2}$$

$\therefore\ 100\,000 \times 1{\cdot}161^{1{\cdot}2} = 119\,618$ (=constant) $= 500\,000 v_2{}^{1{\cdot}2}$
hence $v_2 = 0{\cdot}3036$ m³/kg.

$$\text{work done} = \int_1^2 p\,dv = \int_{1{\cdot}161}^{0{\cdot}3036} \frac{119\,618}{v^{1{\cdot}2}}\,dv = \underline{-178\,613\ \text{J/kg}}$$

The negative sign again indicates that work is done on the gas. To calculate heat supplied it is necessary to know the temperature T_2. Knowing that pv^n = constant and $pv = RT$ $\therefore p = RT/v$, then $(RT/v).v^n$ = constant. $\therefore Tv^{(n-1)}$ = constant, or as $v = RT/p$, $pv^n = p(RT/p)^n$ and hence $T/p^{(n-1)/n}$ = constant.

Thus $T_1 v_1{}^{(n-1)} = T_2 v_2{}^{(n-1)}$.

$$300 \times 1{\cdot}161^{0{\cdot}2} = T_2 \times 0{\cdot}3036^{0{\cdot}2} \quad \therefore\ T_2 = 392{\cdot}31\ \text{K}$$

$$q = (e_2 - e_1) + w = C_v(T_2 - T_1) + w$$

$$= 717{\cdot}9(392{\cdot}31 - 300) - 178\,613$$

$$= \underline{\underline{-112\,343\ \text{J/kg}}}$$

i.e. a heat loss from the system.

THE STEADY FLOW ENERGY EQUATION (SFEE)

All those forms of energy which appear in compressible flow processes are embodied in the steady flow energy equation. Referring to Fig. 68 the principles of conservation of energy (first law) and conservation of mass are applied. At this stage the effects of friction in

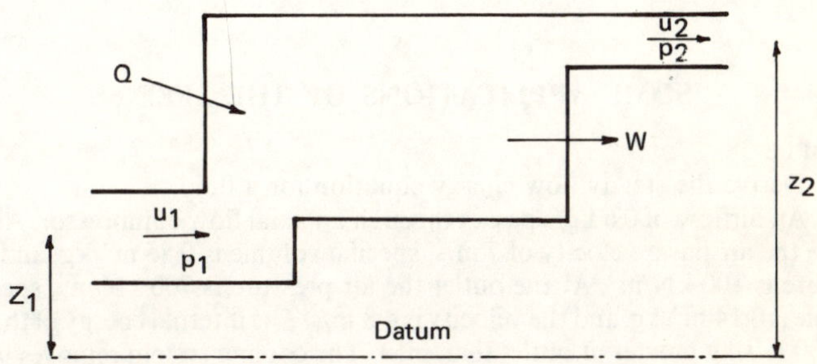

FIG. 68

the system are ignored. The energy entering the system for a mass flow of $\dot{m}$ kg/s is the sum of the kinetic, potential and internal energies: $\frac{1}{2}\dot{m}u_1{}^2 + \dot{m}gz_1 + \dot{m}e_1$. For unit mass flow this is divided by $\dot{m}$, thus

$$\frac{u_1{}^2}{2} + gz_1 + e_1$$

similarly energy leaving the system is

$$\frac{u_2{}^2}{2} + gz_2 + e_2$$

Work done per second due to pressure acting on a fluid in motion is equal to the pressure × volume displaced per second (see p. 41). As volume per second is $\dot{m}/\rho$, work done per second is $p\dot{m}/\rho$ and thus for unit mass, work done per second is p/ρ. Therefore work done on fluid entering system is $+p_1/\rho_1$ and work done on fluid leaving system is $-p_2/\rho_2$. The power output from the system is W, or for unit mass of fluid, $W/\dot{m} = w$. The heat supplied or rejected per second is Q, or for unit mass of fluid $Q/\dot{m} = q$.

From the statement of the first law, the total energy input must equal total energy output, so for frictionless flow

$$\frac{u_1{}^2}{2} + gz_1 + e_1 + \frac{p_1}{\rho_1} + q = \frac{u_2{}^2}{2} + gz_2 + e_2 + \frac{p_2}{\rho_2} + w$$

but as enthalpy, $h = e + p/\rho$, the energy equation becomes

$$\frac{u_1{}^2}{2} + gz_1 + h_1 + q = \frac{u_2{}^2}{2} + gz_2 + h_2 + w$$

This is the *steady flow energy equation*, and the abbreviation "SFEE" is usually employed.

In many practical cases, the change in potential energy is small compared to the other terms and the gz terms are therefore often ignored.

SOME APPLICATIONS OF THE SFEE

EXAMPLE 38

(*a*) Derive the steady flow energy equation for a fluid.

(*b*) An airflow of 0·6 kg/s passes through an axial flow compressor. At the intake the air has a velocity of 7 m/s, specific volume is 0·85 m^3/kg, and the pressure is 100 kN/m^2. At the outlet the air pressure is 700 kN/m^2, specific volume is 0·14 m^3/kg, and the velocity is 4·8 m/s. The internal energy of the air is 18 000 kJ/kg greater at outlet than inlet. The cooling system removes 9000 kJ/s of heat. Calculate the power input to the compressor and the diameter of the intake and exit pipelines, ignore all mechanical or electrical losses in the drive system.

SOLUTION

The first part of the question has already been covered in the preceding text.

For the second part of the question the SFEE is rearranged:

$$-w = \frac{u_2{}^2 - u_1{}^2}{2} + h_2 - h_1 - q$$

Taking each part of the right-hand side of the equation in turn:

$$\frac{u_2{}^2 - u_1{}^2}{2} = \frac{4{\cdot}8^2 - 7{\cdot}0^2}{2} = -26 \text{ J/kg.s}$$

$$\begin{aligned} h_2 - h_1 &= (e_2 - e_1) + (p_2v_2 - p_1v_1) \\ &= 18\,000 + (700\,000 \times 0{\cdot}14 - 100\,000 \times 0{\cdot}85) \\ &= 31\,000 \text{ J/kg.s} \end{aligned}$$

$$q = \frac{Q}{\dot{m}} = -\frac{9000}{0{\cdot}6} = -15\,000 \text{ J/kg.s}$$

substituting in the SFEE gives

$$-w = 31\,000 - 26 - (-15\,000) = 45\,974 \text{ W}$$

$$\underline{\underline{\text{say } 46 \text{ kW}}}$$

If a = cross sectional area of pipeline then $\dot{m} = \rho ua = ua/v$.

$$\therefore \text{ At intake } 0.6 = \frac{7{\cdot}0 \times a}{0{\cdot}85} = \frac{7{\cdot}0 \times \pi d_1{}^2/4}{0{\cdot}85}$$

∴ Diameter at inlet $d_1 = 304{\cdot}77$ mm, say 305 mm.

$$\text{At outlet } 0.6 = \frac{4{\cdot}8 \times a}{0{\cdot}14} = \frac{4{\cdot}8 \times \pi d_2^{\,2}/4}{0{\cdot}14}$$

∴ Diameter at outlet $d_2 = 149{\cdot}27$ mm, say 150 mm.

STAGNATION CONDITIONS

The term "stagnation point" has already been encountered (p. 46) in connection with the use of the pitot tube in an incompressible flow. The same basic idea applies for compressible flows, *i.e.* that a stagnation point is a position on a stream-line at which a flowing fluid is brought to rest. However, with *high speed* gas flows the gas will be compressed as it is brought to rest. The compression process which occurs during deceleration is usually assumed to be isentropic.

As no work is done and no heat is added the SFEE becomes:

$$\frac{u_1^{\,2}}{2} + gz_1 + h_1 = \frac{u_2^{\,2}}{2} + gz_2 + h_2$$

At the stagnation point (position 2) the velocity u_2 will be zero and also the change in potential energy is assumed to be negligible so the gz terms may be omitted. The SFEE thus simplifies to

$$\frac{u_1^{\,2}}{2} + h_1 = h_2$$

For a perfect gas $h = C_pT$, but it is convenient to substitute for C_p in terms of R and γ thus:

$$R = C_p - C_v \quad \text{and} \quad C_p/C_v = \gamma$$

$$\therefore \frac{R}{C_p} = 1 - \frac{C_v}{C_p} = 1 - \frac{1}{\gamma} = \frac{\gamma - 1}{\gamma}$$

$$\therefore C_p = \frac{R\gamma}{\gamma - 1} \qquad \therefore h = \frac{R\gamma}{\gamma - 1}T$$

substituting for h_1 and h_2 in the SFEE gives:

$$\frac{u_1^{\,2}}{2} + \frac{R\gamma}{\gamma - 1}T_1 = \frac{R\gamma}{\gamma - 1}T_2$$

$$\therefore \frac{T_2}{T_1} = 1 + \frac{u_1^{\,2}}{2}\left(\frac{\gamma - 1}{R\gamma T_1}\right) \tag{2}$$

But $T/p^{(n-1)/n} = \text{constant}$ ∴ $T_1/p_1^{\,(n-1)/n} = T_2/p_2^{\,(n-1)/n}$ (see p. 136).

Hence if $n = \gamma$

$$\frac{T_2}{T_1} = \left(\frac{p_2}{p_1}\right)^{(\gamma-1)/\gamma} \tag{3}$$

Now it has been pointed out that compressibility has a significant effect on the stagnation conditions only for high-speed flows. One of the principal parameters of high-speed flow is the Mach number, $M_a = u/c$, where c = speed of sound in the fluid. It is, therefore, usual to relate stagnation temperature and pressure conditions to M_a.

Recalling that the speed of sound in a fluid is given by the equation $c = \sqrt{K/\rho}$ (see p. 59) and also that the bulk modulus $K = -\mathrm{d}p/(\mathrm{d}v/v)$, an expression for c in terms of γ and R may be developed.

As ρ = mass/volume = m/v

$$\therefore \mathrm{d}\rho = \mathrm{d}\left(\frac{m}{v}\right) = -\frac{m\mathrm{d}v}{v^2} = \frac{\rho\mathrm{d}v}{v}$$

$$\therefore \frac{\mathrm{d}\rho}{\rho} = \frac{-\mathrm{d}v}{v}$$

$$\therefore K = \frac{\mathrm{d}p}{(\mathrm{d}\rho/\rho)} = \frac{\rho\mathrm{d}p}{\mathrm{d}\rho}$$

$\therefore$ substituting for K in the speed of sound equation

$$c = \sqrt{\frac{\mathrm{d}p}{\mathrm{d}\rho}}$$

but $\dfrac{p}{\rho^\gamma}$ = constant for an isentropic process

$$\therefore \frac{\mathrm{d}p}{\mathrm{d}\rho} = \frac{\gamma p}{\rho}$$

and as $p/\rho = RT$,

$$\frac{\mathrm{d}p}{\mathrm{d}\rho} = \gamma RT$$

$$\therefore c = \sqrt{\gamma RT} \tag{4}$$

Substituting equations (3) and (4) into equation (2)

$$\frac{T_2}{T_1} = \left(\frac{p_2}{p_1}\right)^{(\gamma-1)/\gamma} = 1 + \frac{u_1{}^2(\gamma-1)}{2c_1{}^2} = 1 + \frac{M_{a1}{}^2(\gamma-1)}{2} \tag{5}$$

It should be noted that as M_a becomes small $M_a{}^2 \to 0$, and therefore

$$\left(\frac{p_2}{p_1}\right)^{(\gamma-1)/\gamma} \to 1$$

Compressibility effects then become negligible and the simpler expressions based on Bernoulli's equation may be used. In practice if M_a is less than about 0·3 it is usual to assume that compressibility may be neglected. Note, also, that for $M_a > 5$, γ is no longer constant.

A typical example of the application of these equations arises in the measurement of velocity of flow of a gas using the Pitot tube.

EXAMPLE 39

A pitot tube inserted into a pipeline carrying CO_2 records a stagnation pressure of 162 kN/m^2 and a static pressure and temperature of 105 kN/m^2 and 25 deg.C respectively. Estimate the velocity of flow. For CO_2, $R = 189$ J/kg.K and $\gamma = 1{\cdot}29$.

SOLUTION

Using equation (5) above,

$$\left(\frac{p_2}{p_1}\right)^{(\gamma-1)/\gamma} = \left(\frac{162\,000}{105\,000}\right)^{0.29/1.29} = 1 + \frac{M_{a1}^{\ 2}}{2}(1{\cdot}29 - 1)$$

$$\therefore\ M_{a1} = 0{\cdot}84$$

Absolute temperature $= 273 + 25 = 298\ \text{K} = T_1$

$$\therefore \text{Speed of sound, } c_1 = \sqrt{\gamma R T_1} = \sqrt{1{\cdot}29 \times 189 \times 298}$$

$$= 269{\cdot}55\ \text{m/s}$$

But $M_{a1} = \dfrac{u_1}{c_1}$ $\qquad \therefore u_1 = c_1 M_{a1}$

$$\therefore u_1 = 269{\cdot}55 \times 0{\cdot}84$$

$$\therefore u_1 = \underline{\underline{226\ \text{m/s}}}$$

NOZZLES

A nozzle is a duct or pipe of smoothly varying cross section, designed to cause a local acceleration to a flowing fluid. For compressible flows there are a number of important applications of nozzles, *e.g.* in steam and gas turbines, jet engines and flow measurement. Fig. 69 shows sections through a converging–diverging nozzle and a converging nozzle.

For nozzles, no work is done, heat loss is assumed to be negligible and z terms are ignored; the SFEE then reduces to:

$$h_1 + \frac{u_1^{\ 2}}{2} = h_2 + \frac{u_2^{\ 2}}{2}$$

or rearranging in terms of u_2,

$$u_2^{\ 2} = 2(h_1 - h_2) - u_1^{\ 2}$$

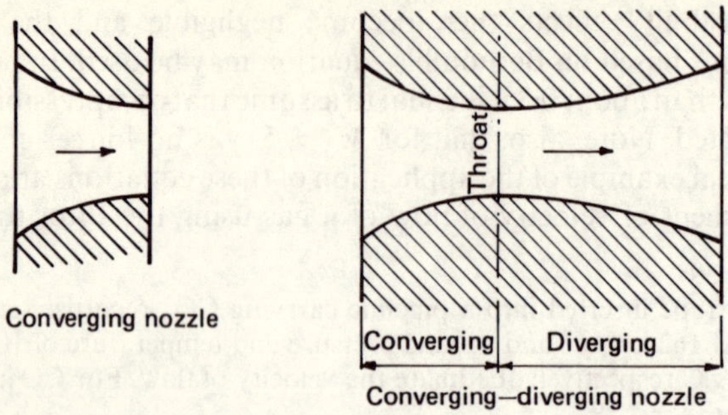

FIG. 69. Nozzles

where station 2 may be at any point along the nozzle.

If velocity at 1 is small compared to velocity at 2,

$$u_2 = \sqrt{2(h_1 - h_2)}$$

This would be true for the case of a large reservoir discharging through a nozzle. For steady flow the mass flow, $\dot{m}$, is constant through the nozzle, and as the cross-sectional area a varies, the ratio $a/\dot{m}$ varies. As $\dot{m} = \rho a u$, $a/\dot{m} = 1/\rho u$. If velocity at 1 is neglected:

$$\frac{a_2}{\dot{m}} = \frac{1}{\rho_2 u_2} = \frac{1}{\rho_2\sqrt{2(h_1 - h_2)}} = \frac{1}{\rho_2\sqrt{2C_p(T_1 - T_2)}}$$

$$= \frac{1}{\rho_2\sqrt{2C_pT_1(1 - T_2/T_1)}}$$

From the equation of state, $p_2 = \rho_2 R T_2$. Also if $p_2/p_1 = \phi$, then for adiabatic flow

$$\frac{T_2}{T_1} = \phi^{(\gamma-1)/\gamma}$$

$\therefore$ substituting $p_2 = \phi p_1$, $T_2 = T_1\phi^{(\gamma-1)/\gamma}$ and $\rho_2 = p_2/RT_2$

$$\frac{a_2}{\dot{m}} = \frac{RT_2}{p_2\sqrt{2C_pT_1(1 - \phi^{(\gamma-1)/\gamma})}} = \frac{RT_1\phi^{(\gamma-1)/\gamma}}{p_1\phi\sqrt{2C_pT_1(1 - \phi^{(\gamma-1)/\gamma})}}$$

This is true anywhere along the nozzle providing flow is adiabatic. The equation is *not* valid, however, for flow downstream of a shock wave (see p. 145) if one occurs within the nozzle.

If conditions $(p_1\rho_1T_1)$ in the reservoir are constant, then the equation may be written:

$$\frac{a_2}{\dot{m}} = \frac{\text{constant}}{\phi^{1/\gamma}\sqrt{1 - \phi^{(\gamma-1)/\gamma}}} = \frac{\text{constant}}{\sqrt{\phi^{2/\gamma} - \phi^{(\gamma+1)/\gamma}}}$$

For maximum flow through a nozzle, $\dot{m}/a \to$ maximum, *i.e.* $a/\dot{m} \to$ minimum, and the denominator of the equation must be a maximum. To determine when this occurs the expression must be differentiated.

$$\frac{d(a/\dot{m})}{d\phi} = -\frac{\left(\frac{2}{\gamma}\right)\phi^{(2/\gamma)-1} - \left(\frac{\gamma+1}{\gamma}\right)\phi^{[(\gamma+1)/\gamma]-1}}{2(\phi^{2/\gamma} - \phi^{(\gamma+1)/\gamma})^{3/2}}$$

$$= 0 \text{ for min}$$

This can be true only if

$$\frac{2}{\gamma}\,\phi^{(2/\gamma)-1} = \left(\frac{\gamma+1}{\gamma}\right)\phi^{[(\gamma+1)/\gamma-1]}$$

i.e.

$$\frac{2}{\gamma+1} = \phi^{(\gamma-1)/\gamma} = \frac{T_2}{T_1} \qquad \text{for maximum flow}$$

The highest value of $\dot{m}/a$ must occur at the throat (narrowest point) of the nozzle. The corresponding velocity at throat is

$$u_t = \sqrt{2(h_1 - h_t)} = \sqrt{2C_p(T_1 - T_t)} = \sqrt{2C_pT_t\left(\frac{T_1 - 1}{T_t}\right)}$$

substituting for

$$C_p = \frac{\gamma R}{\gamma - 1} \quad \text{and} \quad \frac{T_1}{T_t} = \frac{\gamma+1}{2}$$

$$u_t = \sqrt{\frac{2\gamma R}{\gamma - 1}\,T_t\left(\frac{\gamma+1}{2} - 1\right)} = \sqrt{\gamma R T_t} = \text{speed of sound}$$

i.e. for maximum flow in a nozzle the velocity at the throat must be sonic velocity. This is known as the *choked flow* condition. The mass flow rate may be varied between zero flow and choked flow by varying the pressure ratio p_1/p_0, but once the choked condition is reached any further increase in pressure ratio has no effect on flow rate. The physical reason for this is that pressure "signals" of any sort are transmitted through a gas at the speed of sound. Thus for example if p_1 is fixed and p_0 is reduced in small "steps", each reduction is transmitted at sonic velocity back up stream (*i.e.* in the opposite direction to flow) through the nozzle, and the flow rate in the nozzle adjusts itself to suit each new pressure ratio. However, when choked flow is attained the velocity at the throat is sonic velocity, and therefore pressure changes cannot be transmitted upstream past this point. Consequently any further drop in p_0 is not "sensed" upstream of the nozzle throat and the mass flow rate cannot change.

Converging–diverging nozzles are designed for choked flow operation.

Typical effects of the pressure ratio on the variation of pressure and Mach number (or velocity) along the nozzle are illustrated in Fig. 70. Friction has been neglected.

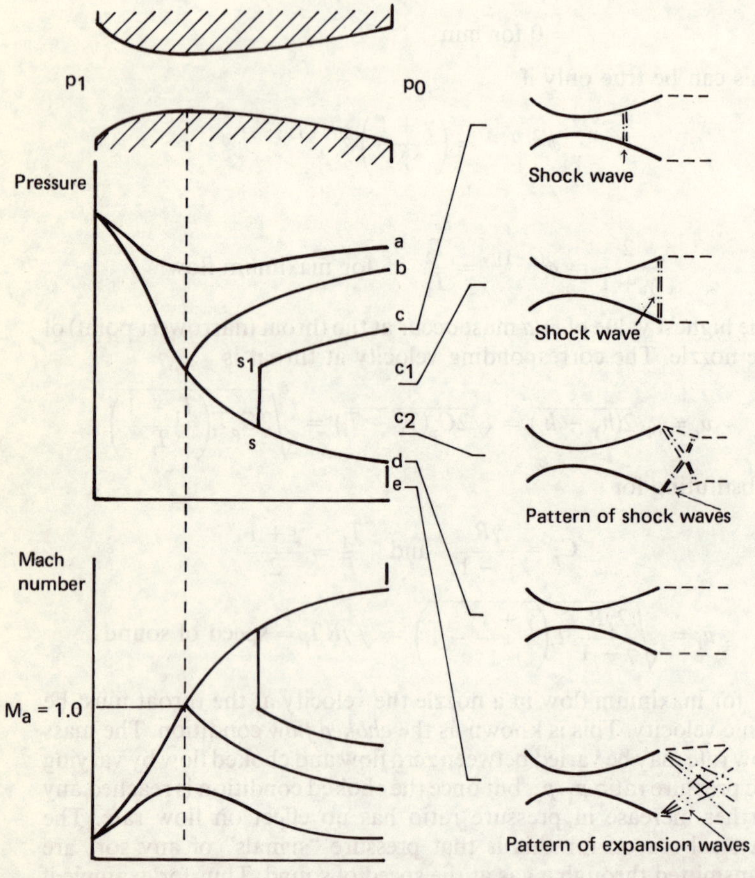

FIG. 70. Variations of pressure and Mach number in a converging–diverging nozzle

i) If p_0 lies between *a* and *b* the pressure ratio p_1/p_0 is insufficient to bring about sonic velocity at the throat. The flow is, therefore, subsonic ($M_a < 1$) throughout the nozzle. Pressure and velocity both vary smoothly through the nozzle, with the gas expanding (p reducing) in the converging section and compressing (p increasing) adiabatically, in the diverging section.

(*ii*) If p_0 is at *b* sonic velocity ($M_a = 1$) is just achieved at the throat, and in the diverging section adiabatic compression takes place.

(*iii*) When p_0 lies between *b* and *d* the flow pattern becomes rather more complex. Choked flow having been attained, the variation of *p* and velocity in the converging section is unaffected by further reductions in p_0. However, in the diverging section the flow accelerates and becomes supersonic with a corresponding expansion. This process cannot continue through to outlet as the pressure must return to p_0 at point *c*. Therefore a region known as a "shock wave" forms at *S*. In a shock wave the flow sharply compresses, and decelerates to sonic velocity at S_1 with a substantial loss of energy. Downstream of S_1 the flow decelerates smoothly (adiabatically) becoming more subsonic. As p_0 is progressively reduced from *c* to c_1 the position of the shock wave moves correspondingly downstream towards the outlet. If p_0 is reduced as far as c_2, the shock wave pattern forms in the jet downstream of the outlet. For $b > p_0 > d$ the flow in the nozzle is said to be "over expanded" since recompression (in the shock wave) has to take place.

(*iv*) If p_0 is at *d*, the flow expands and accelerates smoothly and adiabatically. This is the ideal (or design) pressure ratio.

(*v*) If the pressure just downstream of the outlet is lower than *d*, the flow conditions within the nozzle vary as for (*iv*) but in the jet downstream of the nozzle outlet the pressure has to drop from *d* to *e*. A zone in which a pressure drop occurs in a supersonic flow is known as an "expansion wave".

EXAMPLE 40

A converging–diverging nozzle has a throat diameter of 30 mm and is to operate at choked flow. It is connected to a reservoir of compressed air at a pressure and temperature of 800 kN/m² and 300 K, respectively. The nozzle is to discharge into an atmosphere at 100 kN/m² and 295 K. Estimate (*a*) the mass flow assuming choked flow, (*b*) nozzle diameter at outlet for correct expansion, (*c*) air velocity and Mach number at outlet. For air $R = 287{\cdot}1$ J/kg.K, $\gamma = 1{\cdot}4$, $C_p = 1005$ J/kg.K.

SOLUTION

(*a*) At throat

$$\frac{p_t}{p_1} = \left(\frac{2}{\gamma + 1}\right)^{\gamma/(\gamma - 1)} = \phi = 0{\cdot}528$$

$$\therefore\ \phi^{(\gamma - 1)/\gamma} = 0{\cdot}833$$

$$\frac{a_t}{\dot{m}} = \frac{RT_1\phi^{(\gamma-1)/\gamma}}{p_1\phi\sqrt{2C_pT_1(1 - \phi^{(\gamma-1)/\gamma})}} = \frac{0{\cdot}03^2 \times \pi/4}{\dot{m}}$$

$$= \frac{287{\cdot}1 \times 300 \times 0{\cdot}833}{800\,000 \times 0{\cdot}528\sqrt{2 \times 1005 \times 300 \times 0{\cdot}167}}$$

$$\therefore\ \dot{m} = 1{\cdot}32 \text{ kg/s}$$

(*b*) At outlet

$$\frac{p_0}{p_1} = \frac{100\,000}{800\,000} = \phi = 0{\cdot}125 \quad \therefore\ \phi^{(\gamma-1)/\gamma} = 0{\cdot}552$$

$$\frac{a_0}{\dot{m}} = \frac{a_0}{1{\cdot}32} = \frac{287{\cdot}1 \times 295 \times 0{\cdot}552}{800\,000 \times 0{\cdot}125\sqrt{2 \times 1005 \times 300 \times 0{\cdot}448}}$$

$$\therefore\ a_0 = 0{\cdot}00118 \text{ m}^2 \qquad \therefore\ d_0 = \underline{\underline{38.7 \text{ mm}}}$$

(*c*) At outlet

$$p_0 = \rho_0 R T_0 \quad \therefore\ 100\,000 = \rho_0 \times 287{\cdot}1 \times 295$$

$$\therefore\ \rho_0 = 1{\cdot}18$$

$$\dot{m} = \rho_0 a_0 u_0 \qquad 1{\cdot}32 = 1{\cdot}18 \times 0{\cdot}00118 \times u_0$$

$$\therefore\ u_0 = 948 \text{ m/s}$$

$$c_0 = \sqrt{\gamma R T} = \sqrt{1{\cdot}4 \times 287{\cdot}1 \times 295} = \underline{\underline{344 \text{ m/s}}}$$

$$M_a = \frac{948}{344} = \underline{\underline{2{\cdot}76}}$$

STEADY COMPRESSIBLE FLOW IN PIPES AND DUCTS

Figure 71 shows a section of pipeline of uniform cross-sectional area, a, in which p, ρ and u vary and in which there is a frictional resistance to flow τ_0. The flow is steady.

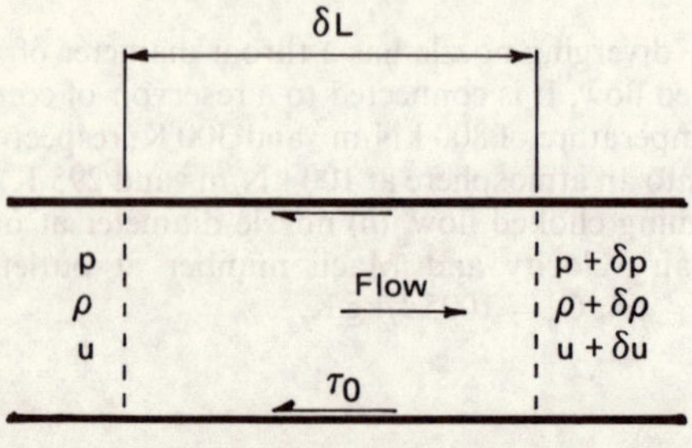

FIG. 71. Compressible flow in pipe

The force applied to the fluid will be the sum of the pressure forces and the frictional resistance. If the frictional shear stress is τ_0, then the frictional forces will be the product of τ_0 and the area of the inner surface of the pipe wall, *i.e.* $\tau_0 \times P\delta L$, where P = inner perimeter of pipe cross section. Because flow is compressible the velocity (and hence momentum) varies along the pipe even though mass flow is constant for steady flow. From the momentum equation:

$$\Sigma \text{ forces} = \text{rate of change of momentum}$$

$$pa - (p + \delta p)a - \tau_0 P\delta L = \dot{m}\delta u = \rho au\delta u$$

$$\therefore \quad -\delta pa - \tau_0 P\delta L = \rho au\delta u$$

Dividing by ρa:

$$-\frac{\delta p}{\rho} - \frac{\tau_0 P\delta L}{\rho a} = u\delta u$$

By definition $\tau_0 = \frac{1}{2}\rho f u^2$ where f = Darcy friction factor.

$$\therefore -\frac{\delta p}{\rho} - \frac{fu^2 P\delta L}{2a} = u\delta u \tag{6}$$

This is a general equation for compressible pipe flow. In order to obtain a solution the compression process must be known.

The most straightforward case is that of isothermal (constant temperature) flow for which p/ρ = constant, and $\therefore \mathrm{d}p/p = \mathrm{d}\rho/\rho$.

As the pipe is of uniform cross section, if the density varies, then the velocity must vary correspondingly to maintain constant mass flow $\dot{m}$ along the pipe. Hence $\dot{m} = \rho au$ = constant, $\therefore \rho u$ = constant.

$$\therefore \frac{\mathrm{d}\rho}{\rho} = -\frac{\mathrm{d}u}{u}$$

$$\therefore \delta p = -p\frac{\delta u}{u} = p\frac{\delta\rho}{\rho} \tag{7}$$

substituting for δp in (6),

$$-\frac{p\delta u}{\rho u} - \frac{fu^2 P\delta L}{2a} = u\delta u$$

dividing by u^2 and recalling that $p/\rho = RT$,

$$-RT\frac{\delta u}{u^3} - \frac{fP\delta L}{2a} = \frac{\delta u}{u}$$

This equation may be solved as a definite integral between stations 1 and 2 distance L apart.

$$\frac{RT}{2}\left(\frac{1}{u_1{}^2} - \frac{1}{u_2{}^2}\right) - \frac{fPL}{2a} = \ln\left(\frac{u_2}{u_1}\right) \tag{8}$$

or as $c = \sqrt{\gamma RT}$ and $M_a = u/c$, equation (8) may be written

$$\frac{1}{2\gamma}\left\{\frac{1}{M_{a_1}^2} - \frac{1}{M_{a_2}^2}\right\} - \frac{fPL}{2a} = \ln\left\{\frac{M_{a_2}}{M_{a_1}}\right\}$$

The maximum flow rate is achieved when the choked flow condition occurs in the pipeline; however, choking may not coincide with sonic velocity (see example). For T to remain constant heat must be transferred to the fluid through the pipe walls. If T varies along the pipe, the relationships between p, ρ, T, and u become more complex.

EXAMPLE 41

Carbon dioxide ($R = 189$, $\gamma = 1{\cdot}304$) is to be conveyed along a 2 km long pipe at a rate of 0·5 kg/s. The pressure and temperature at inlet are 700 kN/m^2 and 35 deg.C, pressure at outlet is 200 kN/m^2 and the friction factor for the pipe is 0·004 (assumed constant). What pipe diameter is required? What is the limiting flow rate for this pipe?

SOLUTION

$p = \rho RT$, $T = 273 + 35 = 308$ deg.K

$$\therefore\ 700\,000 = \rho_1 \times 189 \times 308 \qquad \therefore\ \rho_1 = 12{\cdot}03\ \text{kg/m}^3$$

For isothermal flow p/ρ = constant.

$$\therefore\ \frac{700\,000}{12{\cdot}03} = \frac{200\,000}{\rho_2} \qquad \therefore\ \rho_2 = 3{\cdot}437\ \text{kg/m}^3$$

$\dot{m}$ = constant, pipe cross-sectional area (a) = constant, $\therefore\ \dot{m} = \rho a u$

$$0{\cdot}5 = 12{\cdot}03 \times a \times u_1 = 3{\cdot}437 \times a \times u_2$$

$$\therefore\ u_2 = 3{\cdot}5 \times u_1$$

substituting in equation (8) above for known magnitudes

$$\frac{189 \times 308}{2}\left(\frac{1}{u_1{}^2} - \frac{1}{(3{\cdot}5u_1)^2}\right) - \frac{0{\cdot}004 \times P \times 2000}{2a} = \ln(3{\cdot}5)$$

but $u_1 = \dot{m}/\rho_1 a$, also $a = (\pi/4)d^2$ and $P = \pi d$ (for circular pipe sections).

$$\therefore\ u_1{}^2 = \frac{0{\cdot}0028}{d^4}, \quad \text{and} \quad \frac{P}{a} = \frac{4}{d}$$

substituting for $u_1{}^2$ and P/a gives an equation in which only d is unknown:

$$9\,546\,429d^4 - \frac{16}{d} = 1{\cdot}2528$$

solving by trial and error gives $\underline{\underline{d = 70\ \text{mm}}}$.

To estimate the limiting flow:
From equation (7)

$$-\text{d}u = \frac{u\text{d}p}{p}$$

substituting in equation (6),

$$\frac{\delta p}{\rho} + \frac{fu^2 P\delta L}{2a} = u^2 \frac{\delta p}{p}$$

$$\therefore \frac{\delta p}{\rho}\left(1 - \frac{u^2 \rho}{p}\right) = \frac{fu^2 P\delta L}{2a}$$

$$\therefore \frac{\delta p}{\delta L} = \frac{fu^2 P\rho/(2a)}{(u^2\rho/p) - 1} = \frac{fu^2 P\rho}{2a(\gamma M_a{}^2 - 1)}$$

If $M_a{}^2 \to 1/\gamma$, then $\delta p/\delta L \to \infty$

$\therefore$ Mach number cannot exceed $\sqrt{1/\gamma}$

$$c = \sqrt{\gamma RT} = \sqrt{1{\cdot}304 \times 189 \times 308} = 275{\cdot}5 \text{ m/s}$$

$$\text{maximum Mach number} = \sqrt{1/1{\cdot}304} = 0{\cdot}8757$$

$\therefore$ maximum velocity $u_2 = 0{\cdot}8757 \times 275{\cdot}5 = \underline{\underline{241{\cdot}2 \text{ m/s}}}$

$$\text{actual value of } u_2 = \frac{\dot{m}}{\rho_2 a} = \frac{0{\cdot}5}{3{\cdot}437 \times (\pi/4) \times 0{\cdot}07^2} = \underline{\underline{37{\cdot}8 \text{ m/s}}}$$

$\therefore$ Flow is well below limiting value.

QUESTIONS

For air assume the following values:

$$C_p = 1005 \text{ J/kg.K}; \quad C_v = 718 \text{ J/kg.K}; \quad R = 287; \quad \gamma = 1{\cdot}4$$

$$\rho = 1{\cdot}2 \text{ kg/m}^3 \text{ at } 280 \text{ K and } 101\,500 \text{ N/m}^2 \text{ pressure}$$

1. 1 kg of air initially at 300 K and 100 kN/m^2 is compressed to 500 kN/m^2. Calculate the work done and the heat supplied to the air. Take the value of the polytropic index, n, as 1·2.

2. A convergent–divergent nozzle carries a flow of air and is operating at its design point. Given that the flow is isentropic, prove that:

$$A_r = \sqrt{14{\cdot}93 T_r{}^5(1 - T_r)}$$

where

$$A_r = \frac{\text{throat area}}{\text{outlet area}}$$

$$T_r = \frac{\text{Outlet temperature}}{\text{Stagnation temperature}}$$

3. A pipeline of 90 mm internal diameter incorporates a venturimeter with a throat diameter of 30 mm. The absolute pressure at the inlet to the meter is 130 kN/m^2 whilst at the throat the absolute pressure is 98 kN/m^2. The inlet temperature is 10 deg C and the flow is isentropic. Given that the pipeline carries helium gas, estimate the mass flow.

For helium, $C_p = 5240$ J/kg.K; $C_v = 3157$ J/kg.K.

4. A pitot-static tube is used to measure the velocity in a high-speed wind tunnel. The static pressure is 146.2 kN/m^2 (absolute), while the stagnation temperature and pressure are 370 K and 250 kN/m^2 (absolute). Calculate

(*a*) the true air velocity assuming compressible flow;
(*b*) the percentage error if the velocity is calculated using Bernoulli's equation.

5. A converging–diverging nozzle has a throat area of 6 cm^2 and an outlet area of 10 cm^2. The nozzle is fixed to the side of a large tank which contains carbon dioxide at a pressure and temperature of 600 kN/m^2 (absolute) and 18 deg.C. Calculate the design flow rate and the design pressure at outlet.
For CO_2, $C_p = 834$ J/kg.K, $C_v = 572$ J/kg.K.

6. Air flows along a horizontal duct which is 300 m long. The duct is of square cross section, 150 mm × 150 mm. For a mass flow rate of 5.73 kg/s the inlet pressure is 500 kN/m^2, the outlet pressure is 125 kN/m^2 and the air temperature is constant at 57 deg.C. Assuming that the flow is free of shock waves, calculate:

(*a*) the friction factor for the duct;
(*b*) the rate of heat transfer required to maintain the air temperature along the duct;
(*c*) the Mach number of the flow at outlet.

CHAPTER 10

BOUNDARY LAYER THEORY

AN *ideal* fluid is incompressible, does not vaporise, has no surface tension and has no viscosity. Many fluids, particularly water, behave as *ideal* wherever they are remote from the influence of solid boundaries and the flow can then by analysed using the concepts of theoretical hydrodynamics and ignoring fluid friction. Once the flow *is* influenced by solid boundaries, however, the laws governing it, particularly near the boundaries, are very different. Prandtl first investigated the problem and subsequently explained the changes by enunciating the *Boundary Layer Theory* at the beginning of this century.

The theory can best be explained by considering flow past a flat plate situated in an otherwise uniform stream. The plate and boundary layer is illustrated in Fig. 72 (*a*), but it should be appreciated that to make the illustration meaningful, the vertical scale is greatly enlarged.

MAIN FEATURES OF A BOUNDARY LAYER

If a flat plate were set up in a uniform stream as shown in Fig. 72 (*a*) and if it were possible to take appropriate measurements with

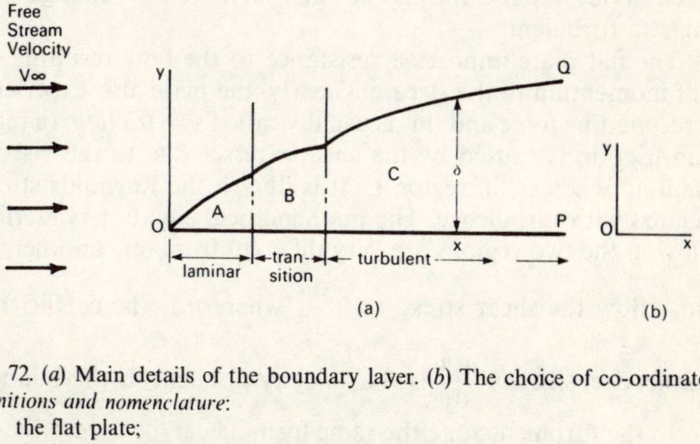

FIG. 72. (*a*) Main details of the boundary layer. (*b*) The choice of co-ordinate axes

Definitions and nomenclature:

OP: the flat plate;
O: the leading edge and the origin of the x–y system;
OQ: the extent of the boundary layer;
V_∞: the free stream velocity = the uniform velocity of the fluid everywhere except within the boundary layer;
δ: thickness of the boundary layer at a distance x from *O*.

sufficient accuracy, the boundary layer would be found to exhibit the features listed below.

(*a*) A boundary layer would develop both sides of the plate. In Fig. 72 (*a*) it is shown on one side only of the plate in order to simplify the illustration.

(*b*) The velocity of the fluid in the boundary layer is zero at the plate and increases to the maximum value V_∞ on OQ. The laws governing the velocity distribution across the boundary layer are different for the two regions A and C. For all practical purposes the velocity in the boundary layer may be considered to be parallel to the plate.

(*c*) The thickness of the boundary layer is very small compared with its length and for this reason the y-scale has had to be greatly exaggerated to give a reasonably proportioned illustration.

(*d*) Three distinct regions, A, B and C would develop. In A the flow would be laminar, in B it would be transitional between A and C and in C it would be turbulent.

(*e*) In most normal flow problems the extent of region A, in which the flow is laminar, is very short and region B is usually shorter. For many practical problems, therefore, the flow can be thought of as turbulent throughout the whole boundary layer with little loss of accuracy; although this is strictly true only if the thickness of the boundary layer is very small compared with the length, L, of the plate.

(*f*) A Reynolds number can be defined which has a critical value between about 4×10^5 and 2×10^6 at which the flow changes from laminar to turbulent.

(*g*) The flat plate imposes a resistance to the flow resulting in a loss of momentum of the stream. Clearly, the plate also experiences a corresponding force and this is usually called *skin friction*. In region A skin friction is caused by the shear stresses due to the viscosity of the fluid whereas, in region C, it is due to the Reynolds stresses associated with turbulence. The mathematical equations describing the flow in the two regions are very different from one another. For laminar flow the shear stress $\tau = \mu \dfrac{du}{dy}$ where μ is the coefficient of dynamic viscosity and $\dfrac{du}{dy}$ is the velocity gradient in the boundary layer. In the turbulent zone the same formula can be used providing μ is re-defined as *eddy viscosity*.

(*h*) The boundary layer grows in thickness as the distance downstream from O increases; *i.e.* δ increases with x.

(*i*) If the plate were smooth it would be found that a *laminar sublayer* forms between the plate and the turbulent flow in region C; it is too thin to illustrate on the sketch. If the plate is not perfectly smooth the laminar sublayer may still form depending on the degree of the roughness. For example, in experiments roughness is often artificially produced by sticking sand grains to the surface; then, if the diameter of the grains is less than the thickness of the laminar sublayer, it will still form and the plate will behave as though it were smooth. If, however, the roughness (or grain diameter) is greater than the thickness of the sublayer, so that it protrudes through, the high points will shed eddies and the plate may be considered rough. The laminar sublayer becomes thinner as R_e is increased. Consequently, every plate can behave either as smooth or rough depending on the Reynolds number.

(*j*) Although the ensuing bookwork is based on the case of a flat plate in a uniform stream, the applications of boundary layer theory are very much wider and are briefly described on pp. 172ff. Particular mention is made here, however, of pipe-flow because, in effect, a pipe can be thought of as a flat plate wrapped round to re-join itself, and much of the empirical data featuring in flat-plate theory is derived from experiments on pipe flow.

(*k*) In boundary layer theory it is normal to make the assumptions that:

(*i*) the pressure is uniform throughout the boundary layer; and
(*ii*) the fluid is assumed to behave as incompressible.

BOUNDARY LAYER THICKNESS

It must be remembered that δ increases as x increases and when boundary layer thickness is spoken of, therefore, it implies a particular value of δ corresponding to one value of x (which may not be specified).

The velocity distribution across a boundary layer cannot be measured to a high degree of accuracy because the boundary layer is so thin and, because the velocity in the boundary layer approaches its maximum value V_∞ asymptotically, the exact value of δ at any given x-value is ill-defined. It is, therefore, sometimes arbitrarily defined as the distance from the boundary to the point where the velocity in the boundary layer reaches 99 per cent of its peak value, V_∞ and it is shown and labelled as δ_{99} in Fig. 73.

Three other measurements related to boundary layer thickness are the *displacement thickness*, *energy thickness* and *momentum thickness*. The velocity of flow, u, in the boundary layer is everywhere less than the free stream velocity, V_∞, and it follows that, in the presence of a flat plate, (*a*) the main stream is displaced slightly by the boundary layer, (*b*) there is a net loss of kinetic energy in the system and (*c*) there is a net loss of momentum. The three definitions are given in Example 42.

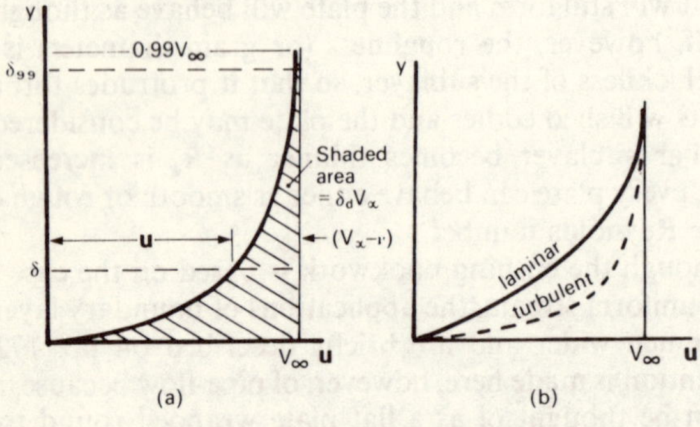

FIG. 73. (*a*) Velocity distribution for turbulent flow in a boundary layer δ_d and δ_{99} are parameters for boundary-layer thickness (*b*) Velocity distribution in a laminar boundary layer with the "turbulent" velocity distribution shown dotted for comparison

EXAMPLE 42

(*a*) Describe the characteristics of a boundary layer formed on a smooth flat plate placed parallel to the direction of flow of a uniform stream of incompressible fluid. Assume the pressure is uniform and that the plate is stationary.

(*b*) Explain the concept and use of the *displacement thickness* δ_d, the *energy thickness*, δ_e, and the *momentum thickness*, δ_m.

SOLUTION

(*a*) The solution to this part is given in the preceding pages.

(*b*) If the free stream velocity is V_∞, the velocity at a distance of y from the boundary is u and is assumed to be parallel to the plate, and the boundary layer thickness is δ, then:

displacement thickness δ_d, is defined by the equation,

$$\delta_d = \int_0^\delta \left(1 - \frac{u}{V_\infty}\right) \mathrm{d}y$$

and represents the amount by which the streamlines in the potential flow outside the boundary layer are displaced owing to the reduced velocity within the boundary layer. In practical situations the velocity and depth of the uniform stream past a flat plate can therefore be calculated by treating it as though it was inviscid flow past a solid body of thickness δ_d.

Energy thickness, δ_e, is defined by the equation,

$$\delta_e = \int_0^\delta \frac{u}{V_\infty}\left[1 - \left(\frac{u}{V_\infty}\right)^2\right] \mathrm{d}y$$

and is such that the loss of kinetic energy due to the presence of the boundary layer is equivalent to a reduction of δ_e in the total depth of the uniform stream; *i.e.* the loss of flux of K.E. per unit width of boundary layer is $\frac{1}{2}\rho V_\infty{}^3\delta_e$, where ρ = density of the fluid.

Momentum thickness δ_m, is defined by the equation,

$$\delta_m = \int_0^\delta \frac{u}{V_\infty}\left(1 - \frac{u}{V_\infty}\right) \mathrm{d}y$$

and is such that the loss of rate of change of momentum due to the boundary layer is equivalent to a reduction of δ_m in the total depth of the uniform stream; *i.e.* owing to the presence of the boundary layer the oncoming uniform stream suffers a loss of momentum given by $\rho V_\infty{}^2\delta_m$. Since *force = rate of change of momentum*, the force on the plate due to skin friction is also given by $\rho V_\infty{}^2\delta_m$.

EXAMPLE 43

(*a*) Determine the relationship between δ and δ_d for a boundary layer which is (*i*) laminar throughout and (*ii*) turbulent throughout. Assume:

(*i*) in the laminar boundary layer, the flow obeys the law

shear stress, $\tau = \mu\dfrac{\mathrm{d}u}{\mathrm{d}y}$, where μ is the viscosity; and

(*ii*) the velocity distribution in the turbulent boundary layer is given by:

$$\frac{u}{V_\infty} = \left(\frac{y}{\delta}\right)^{1/7}$$

(*b*) In tests in a wind tunnel measuring 600 mm square in section the boundary layer thickness on each side of a flat plate is estimated to be 50 mm. The fluid is compressed air which may be assumed to be of constant density, the upstream uniform velocity, V_∞ is 3 m/s and the boundary layer is laminar. If the pressure distribution on the plate is uniform, calculate the changed value of V_∞ near the plate.

SOLUTION

(*a*) (*i*) Assuming $\tau \propto y$ and μ is constant, $\dfrac{\mathrm{d}u}{\mathrm{d}y} \propto y$ and this implies that the velocity distribution is parabolic because no other curve has this property. The equation to the velocity profile may, therefore, be written:

$$(V_\infty - u) = k(\delta - y)^2$$

and since $u = 0$, when $y = 0$, the constant is $k = V_\infty/\delta^2$. Hence:

$$(V_\infty - u) = \frac{V_\infty}{\delta^2}(\delta - y)^2$$

The displacement thickness is defined by $\delta_d = \frac{1}{V_\infty}\int_0^\delta (V_\infty - u)\mathrm{d}y$

Hence $$\delta_d = \frac{1}{V_\infty}\int_0^\delta \frac{V_\infty}{\delta^2}(\delta - y)^2\,\mathrm{d}y = \int_0^\delta \left(1 - \frac{y}{\delta}\right)^2 \mathrm{d}y$$

Substituting $r = \frac{y}{\delta}$: $\mathrm{d}y = \delta \mathrm{d}r$ and the upper limit becomes 1.

$$\therefore \quad \delta_d = \int_0^1 (1 - r)^2 \delta\,\mathrm{d}r$$

$$= \left[-\frac{(1 - r)^3\delta}{3}\right]_0^1 = \frac{\delta}{3}$$

Hence, for *laminar* boundary layers, the displacement thickness is 1/3 the thickness of the layer.

(*ii*) In the *turbulent* region the assumption that $\frac{u}{V_\infty} = \left(\frac{y}{\delta}\right)^{1/7}$ is often made.

Then, putting $\frac{y}{\delta} = r$ and using the same procedure as for the laminar case,

$$\delta_d = \int_0^\delta \left\{1 - \left(\frac{y}{\delta}\right)^{1/7}\right\}\mathrm{d}y = \int_0^1 (1 - r^{1/7})\delta \mathrm{d}r$$

from which, $$\delta_d = \frac{\delta}{8}$$

Both these results in practice must be considered approximate. The result for the turbulent region, moreover, applies only throughout a limited range of the Reynolds number, R_e, for this kind of flow, *see* p. 164. For higher values of R_e the factor 8 increases.

(*b*) Since the boundary layer is laminar, $\delta_d = \frac{1}{3}\delta$.

$$2\delta = 2 \times 50 = 100 \text{ mm}$$

$\therefore$ effective depth of the wind tunnel $= 600 - \frac{100}{3} = 567$ mm.

$\therefore$ locally, V_∞ is increased to $3 \times 600/567 = 3{\cdot}18$ m/s.

MOMENTUM EQUATION FOR THE BOUNDARY LAYER

The nature and characteristics of the boundary layer have been described in the previous pages and the reader will appreciate that the growth of the boundary layer downstream must be accompanied by a slowing down of the flow within it. The only forces acting on the boundary layer are the skin friction and forces caused by pressure variations in the flow above the plate. In this section it is assumed that the pressure on the plate is uniform and the only force to be considered, therefore, is the skin friction. This is equated to the rate of change of momentum of the fluid passing through the region *NOPQ*, *see* Fig. 74 (*a*). *OP* lies along the flat plate, *PQ* is the boundary layer thickness, δ, at a distance x from *O*, *ON* is the distance along the y-axis such that the flow across it (measured in, for example, m^2/s per metre width) is the same as the flow across *PQ* and the curved line *NQ* is the streamline joining *N* and *Q*; *i.e.*, there is no component of flow across it. Furthermore, the flow considered is per unit width of plate.

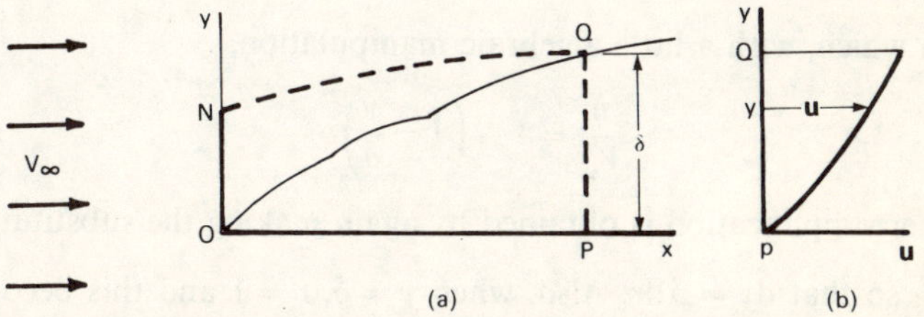

FIG. 74. (*a*) The control volume for the application of the momentum principle. (*b*) The velocity profile at *PQ*

The rate of flow across *PQ* is: $\int_0^\delta u\,\mathrm{d}y$

$\therefore$ rate of change of momentum at $PQ = \int_0^\delta \rho u^2\,\mathrm{d}y$

Since the same rate of flow occurs at *ON* and the velocity there is V_∞, rate of change of momentum at $ON = \int_0^\delta \rho V_\infty u\,\mathrm{d}y$. The boundary layer is very thin and consequently the assumption can again be made that u, which varies throughout the boundary layer, is everywhere parallel to *OP*. As the only force acting on the plate is the skin friction, F (per unit width), it must equal the momentum lost, *i.e.*

$$F = \int_0^\delta \rho V_\infty u\,\mathrm{d}y - \int_0^\delta \rho u^2\,\mathrm{d}y = \rho\int_0^\delta (uV_\infty - u^2)\,\mathrm{d}y = \rho\int_0^\delta u(V_\infty - u)\,\mathrm{d}y$$

or
$$F = \rho V_\infty{}^2 \int_0^\delta \left(\frac{u}{V_\infty} - \frac{u^2}{V_\infty{}^2}\right)\mathrm{d}y$$

This important equation is the *momentum equation* for the boundary layer. To solve it, $\frac{u}{V_\infty}$ has to be expressed in terms of y. This is done below for laminar and turbulent boundary layers, in turn, and expressions for F obtained. It should also be noted that $F = \rho V_\infty{}^2 \delta_m$ where δ_m = momentum thickness.

Laminar boundary layer

In this case use is made of the velocity profile equation given on p. 155, namely

$$(V_\infty - u) = \frac{V_\infty}{\delta^2}(\delta - y)^2$$

from which, with a little algebraic manipulation,

$$\frac{u}{V_\infty} = 1 - \left(1 - \frac{y}{\delta}\right)^2$$

further simplification is obtained by again making the substitution, $r = \frac{y}{\delta}$ so that $\mathrm{d}y = \delta\,\mathrm{d}r$. Also, when $y = \delta$, $r = 1$ and this becomes the upper limit in the integral. Hence:

$$\frac{u}{V_\infty} = 1 - (1 - r)^2 = r(2 - r)$$

$$F = \rho V_\infty{}^2 \int_0^\delta \frac{u}{V_\infty}\left(1 - \frac{u}{V_\infty}\right)\mathrm{d}y$$

$$= \rho V_\infty{}^2 \int_0^1 \{r(2-r) - r^2(2-r)^2\}\delta\,\mathrm{d}r$$

$$= \rho V_\infty{}^2 \delta\left[r^2 - \frac{5}{3}r^3 + r^4 - \frac{r^5}{5}\right]_0^1$$

i.e. $$F = \frac{2}{15}\rho V_\infty{}^2 \delta$$

Turbulent boundary layer

Experiments show that, for Reynolds numbers between about 5×10^6 and 2×10^7, the velocity distribution can be represented with reasonable accuracy by the one-seventh power rule, namely:

$$\frac{u}{V_\infty} = \left(\frac{y}{\delta}\right)^{1/7} = r^{1/7} \qquad \text{where } r = y/\delta$$

Hence,
$$F = \rho V_\infty^2 \int_0^\delta \frac{u}{V_\infty}\left(1 - \frac{u}{V_\infty}\right)\mathrm{d}y$$
$$= \rho V_\infty^2 \int_0^1 (r^{1/7} - r^{2/7})\delta\,\mathrm{d}r$$
$$= \rho V_\infty^2 \delta\left[\frac{7}{8}r^{8/7} - \frac{7}{9}r^{9/7}\right]_0^1$$

i.e.
$$F = \frac{7}{72}\rho V_\infty^2 \delta$$

EQUATION TO BOUNDARY LAYER SURFACES

Another expression may be obtained for the friction force, F, by summing the values of τ_0 over the whole of the flat plate, thus

$$F = \int_0^x \tau_0\,\mathrm{d}x$$

τ_0 is the shear stress developed in the fluid at $y = 0$, *i.e.* at the boundary, and may be represented by the equation

$$\tau_0 = \mu\left\{\frac{\mathrm{d}u}{\mathrm{d}y}\right\}_{y=0}$$

where $\left\{\frac{\mathrm{d}u}{\mathrm{d}y}\right\}_{y=0}$ is the velocity gradient at the boundary (when $y = 0$) and, for laminar boundary layers, μ is the viscosity. The same equation for τ_0 can be used for turbulent boundary layers providing it is remembered that, in this case, μ must be interpreted as the *eddy* viscosity. In both cases the thickness of the boundary layer grows with distance from the leading edge (x) and, as it thickens, the velocity, and therefore $\frac{\mathrm{d}u}{\mathrm{d}y}$, reduces. It follows that τ_0 is a function of x although it is not the same function for both laminar and turbulent cases.

It should also be noted that since $F = \int_0^x \tau_0\,\mathrm{d}x$, $\tau_0 = \frac{\mathrm{d}F}{\mathrm{d}x}$.

Laminar boundary layer

From p. 158
$$F = \frac{2}{15}\rho V_\infty^2 \delta$$

$\therefore$
$$\tau_0 = \frac{\mathrm{d}F}{\mathrm{d}x} = \frac{2}{15}\rho V_\infty^2 \frac{\mathrm{d}\delta}{\mathrm{d}x}$$

A second equation for τ_0 can be obtained from $\tau_0 = \mu\left\{\frac{du}{dy}\right\}_{y=0}$

From p. 158, $$\frac{u}{V_\infty} = r(2 - r) \qquad \text{where } r = \frac{y}{\delta}$$

$$\therefore \qquad \frac{du}{dr} = 2V_\infty(1 - r)$$

and $$\frac{du}{dy} = \frac{du}{dr} \times \frac{dr}{dy} = 2V_\infty(1 - r) \times \frac{1}{\delta}$$

Since $r = 0$ when $y = 0$, $\left\{\frac{du}{dy}\right\}_{y=0} = \frac{2V_\infty}{\delta}$

$$\therefore \qquad \tau_0 = \frac{2\mu V_\infty}{\delta}$$

Equating the two expressions for τ_0:

$$\frac{2\mu V_\infty}{\delta} = \frac{2}{15}\rho V_\infty^{\ 2}\frac{d\delta}{dx}$$

Separating variables, $$\int \frac{30\mu}{\rho V_\infty} dx = \int 2\delta \, d\delta$$

$$\therefore \qquad \frac{30\mu x}{\rho V_\infty} = \delta^2 + C$$

Since $\delta = 0$ when $x = 0$, the constant of integration, $C = 0$.

Hence, $$\left(\frac{\delta}{x}\right)^2 = \frac{30\nu}{x V_\infty} \qquad \text{where } \nu = \frac{\mu}{\rho}$$

i.e. $$\frac{\delta}{x} = \frac{5{\cdot}48}{\sqrt{R_e}} \qquad \text{where } R_e = \frac{xV_\infty}{\nu}$$

which is the theoretical equation to the surface of the laminar boundary layer.

Turbulent boundary layer

The equation $\tau_0 = \mu\left\{\frac{du}{dy}\right\}_{y=0}$ which was used in the laminar case is of no help here because, if the velocity distribution is $\frac{u}{V_\infty} = \left(\frac{y}{\delta}\right)^{1/7}$, $\left\{\frac{du}{dy}\right\}_{y=0} = 0$ and the mathematics breaks down. Recourse has to be made to the results of experiments on pipe flow and these lead to:

$$\tau_0 = 0{\cdot}023\rho V_\infty{}^2\left(\frac{\nu}{V_\infty\delta}\right)^{\frac{1}{4}}$$

From p. 159,

$$F = \frac{7}{72}\rho V_\infty{}^2\delta$$

$$\therefore \qquad \tau_0 = \frac{7}{72}\rho V_\infty{}^2\frac{\mathrm{d}\delta}{\mathrm{d}x}$$

$$\therefore \qquad \int \delta^{\frac{1}{4}}\mathrm{d}\delta = \frac{0{\cdot}023 \times 72}{7}\left(\frac{\nu}{V_\infty}\right)^{\frac{1}{4}}x \qquad \text{where } \nu = \frac{\mu}{\rho}$$

$$\therefore \qquad \frac{4}{5}\delta^{\frac{5}{4}} = 0{\cdot}237\left(\frac{\nu}{V_\infty}\right)^{\frac{1}{4}}x$$

$$\therefore \qquad \delta = 0{\cdot}377x^{\frac{4}{5}}\left(\frac{\nu}{V_\infty}\right)^{\frac{1}{5}} = 0{\cdot}377\left(\frac{\nu}{V_\infty}\right)^{\frac{1}{5}}x$$

$$\text{i.e.} \qquad \frac{\delta}{x} = 0.377R_e{}^{-\frac{1}{5}}$$

which is the equation to the surface of the *turbulent* boundary layer. Whereas this expression has practical value, it should be noted, the theoretical value obtained for the thickness of the *laminar* boundary layer is shown by experimental evidence to be too high and the value used in practice is, therefore,

$$\frac{\delta}{x} = 5{\cdot}00R_e{}^{-\frac{1}{2}}$$

This is the value used in the following example.

EXAMPLE 44

Plot the boundary layer on one side of a flat plate of length L if $R_e = \dfrac{V_\infty L}{\nu} = 10^6$. Assume that the boundary layer is (*a*) laminar throughout and (*b*) turbulent throughout. Comment on the curves obtained.

SOLUTION

The Reynolds number for flow past a flat plate is defined as $\dfrac{V_\infty x}{\nu}$ and reaches its maximum value of $\dfrac{V_\infty L}{\nu}$ only at the trailing edge of the plate.

In Table III the calculated values for R_e are based on x, the distance along the plate. Both x and δ are expressed as a decimal of L.

x	$R_e = \frac{V_\infty x}{\nu}$	"*laminar*" $\delta = 5xR_e^{-\frac{1}{2}}$	"*turbulent*" $\delta = 0{\cdot}377xR_e^{-\frac{1}{5}}$
0·1 L	10^5	0·00158 L	0·00377 L
0·2 L	2×10^5	0·00224 L	0·00656 L
0·3 L	3×10^5	0·00274 L	0·00908 L
0·4 L	4×10^5	0·00316 L	0·01143 L
0·5 L	5×10^5	0·00354 L	0·01366 L
0·7 L	7×10^5	0·00418 L	0·01788 L
L	10^6	0·00500 L	0·02379 L

TABLE III.

The plot is shown in Fig. 75 and illustrates that the thickness of the boundary layer is greater for turbulent boundary layers than for laminar.

In this example the assumption of a *turbulent* boundary layer throughout is artificial because, of course, turbulence would occur over, at most, a relatively short length of the plate at the given Reynold's number. Again it is emphasised that the vertical scale in Fig. 75 is greatly exaggerated.

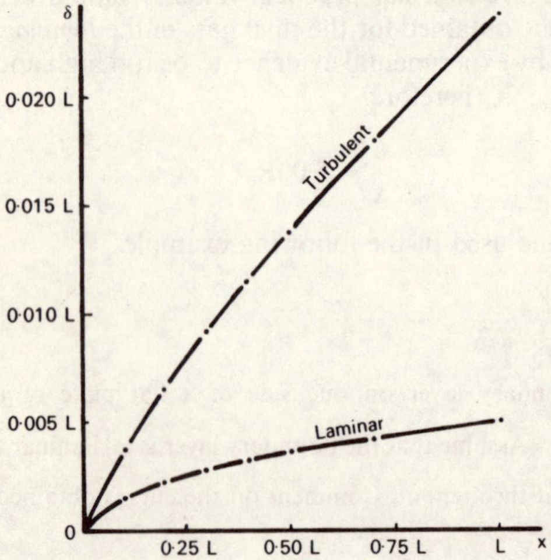

FIG. 75. *See* Example 44. Boundary-layer thickness plotted with greatly magnified vertical scale (50-fold)

COEFFICIENT OF FRICTION, C_f

This is defined as:

$$C_f = \frac{\text{friction force}}{\frac{1}{2}\rho V_\infty{}^2 \times \text{wetted area}} = \frac{F}{\frac{1}{2}\rho V_\infty{}^2 L}$$

For a *laminar* boundary layer the theoretical value is

$$C_f = \frac{2}{15}\rho V_\infty{}^2 \frac{5{\cdot}48L}{\sqrt{R_e}} \div \tfrac{1}{2}\rho V_\infty{}^2 L$$

$$= 1{\cdot}46 R_e{}^{-\frac{1}{2}}$$

In practice it is found that a more correct value is $\underline{\underline{C_f = 1{\cdot}328 R_e{}^{-\frac{1}{2}}}}$.

For a *turbulent* boundary layer,

$$C_f = \frac{7}{72}\rho V_\infty{}^2 \times \frac{0{\cdot}377L}{R_e^{\frac{1}{5}}} \div \tfrac{1}{2}\rho V_\infty{}^2 L$$

i.e. $\quad C_f = 0{\cdot}073\, R_e{}^{-\frac{1}{5}}$

EXAMPLE 45

(*a*) Define a Reynolds number for flow past a flat plate situated in an otherwise uniform stream and explain its relevance to the coefficient of drag. Where formulae are quoted indicate the range of Reynolds numbers over which they are valid.

(*b*) Determine the drag on a flat plate measuring 750 mm × 100 mm submerged in a uniform stream of (*i*) water and (*ii*) air assuming the boundary layer which forms to be either *all* laminar or *all* turbulent. The plate is situated with its longer sides parallel to the flow and the free stream velocity is 3 m/s.

For water, $\rho = 994$ kg/m³ and $\nu = 0{\cdot}684 \times 10^{-6}$ m²/s.
For air, $\rho = 1{\cdot}136$ kg/m³ and $\nu = 0{\cdot}167 \times 10^{-4}$ m²/s.
Critical Reynolds number = 5×10^5.

SOLUTION

(*a*) A Reynolds number for flow past a flat plate is defined as $R_e = \dfrac{V_\infty x}{\nu}$

in which V_∞ = free stream velocity, ν = kinematic viscosity ($= \mu/\rho$) of the fluid and x is the distance from the leading edge to the point on the plate under consideration. For a given situation V_∞ and ν are fixed by the circumstances and, therefore, $R_e \propto x$.

If the plate is of length L, the maximum value of R_e occurs at the downstream end where $x = L$ and is $R_e = \dfrac{V_\infty L}{\nu}$

The Reynolds number is dimensionless and is a measure of the relative importance of the viscosity, being high when ν is small and relatively low when v is large. Laminar boundary layers, in which the flow is viscous, are characterised by the lower Reynold's numbers. In fact, the lower critical value of R_e is about 4×10^5 and the change to turbulent flow occurs usually between this value and $R_e = 2 \times 10^6$.

Drag force is often described as $F = \frac{1}{2}C_f \rho V_\infty{}^2 A$ where C_f is the coefficient of friction, ρ is the fluid density, V_∞ is the free stream velocity and A is the surface area of one side of the plate in contact with the fluid. If the plate is submerged a boundary layer will form on both sides of it and the total drag force will then be $2F$.

C_f can be defined as a function of R_e but the function is not the same for each of the regions A, B and C in Fig. 72, the main functions being as follows: $R_e < 4 \times 10^5$, $C_f = \dfrac{1{\cdot}328}{R_e^{\frac{1}{2}}}$ and the flow is laminar; $5 \times 10^5 < R_e < 2 \times 10^7$, $C_f = \dfrac{0{\cdot}073}{R_e^{\frac{1}{5}}}$ and the flow may be considered to be turbulent throughout although, strictly, in the lower part of this range of Reynolds numbers, the flow may be transitional between laminar and turbulent in that it may be either, a mixture of both or a changing pattern which fluctuates between laminar and turbulent.

$R_e > 2 \times 10^7$ the flow is still turbulent but the velocity distribution is more accurately represented by a logarithmic function which leads to a logarithmic equation for C_f. The laminar sublayer becomes of greater significance too. If its thickness is greater than the roughness, k, (a linear quantity) the plate behaves as hydraulically smooth and $C_f = 0{\cdot}455(\log_{10} R_e)^{-2{\cdot}58}$ whereas if the roughness is greater than the thickness of the sublayer, C_f becomes a function, not of R_e, but of the relative roughness, k/L.

(*b*) The Reynolds number for the flow is $R_e = \dfrac{V_\infty L}{v} = \dfrac{3 \times 0{\cdot}75}{v} = \dfrac{2{\cdot}25}{v}$

For water, $v = 0{\cdot}0684 \times 10^{-5}$ and, therefore, $R_e = 32{\cdot}9 \times 10^5$

For air, $v = 0{\cdot}167 \times 10^{-4}$ and, therefore, $R_e = 13{\cdot}45 \times 10^4$

For water, $C_f = 0{\cdot}073/R_e{}^{1/5} = 0{\cdot}00363$

$$\therefore \quad \text{total drag} = 2F = 2 \times \tfrac{1}{2} \times \rho \times V_\infty{}^2 \times 0{\cdot}75 \times 0{\cdot}100 \times 0{\cdot}00363$$

$$= \underline{\underline{2{\cdot}44\ \text{N}}}$$

For air, $C_f = 1{\cdot}328/R^{\frac{1}{2}} = 0{\cdot}00362$

$$\therefore \quad \text{total drag} = 2F = 2 \times \tfrac{1}{2} \times 1{\cdot}136 \times 9 \times 0{\cdot}75 \times 0{\cdot}1 \times 0{\cdot}00362$$

$$= \underline{\underline{0{\cdot}003\ \text{N}}}$$

DOWNSTREAM VARIATIONS OF PRESSURE

The word *wake* is usually associated with the vortex flow downstream of a *bluff* (or blunt) body but it is convenient here to extend its meaning to include also the region of slower flow which follows a flat plate, *i.e.* where $u < V_\infty$. More is written in the next chapter about the characteristics of a wake and the way in which it forms.

Its importance here is that a sure way of measuring the change of momentum in a stream obstructed by a body is by carrying out a velocity traverse across the wake using, for example, a pitot tube to determine the velocity distribution. Then for two-dimensional flow patterns the rate of change of momentum is $\int_{y_1}^{y_2} \rho u(V_\infty - u)\mathrm{d}y$ where u is the velocity at a distance y measured normal to the wake from a convenient reference point, say the centre of the wake. If y_1 and y_2 are the outer limits of the wake, *i.e.* where $u = V_\infty$, then, in accordance with the principle of conservation of momentum,

$$\int_{y_1}^{y_2} \rho u(V_\infty - u)\mathrm{d}y = \text{sum of the forces acting on the stream.}$$

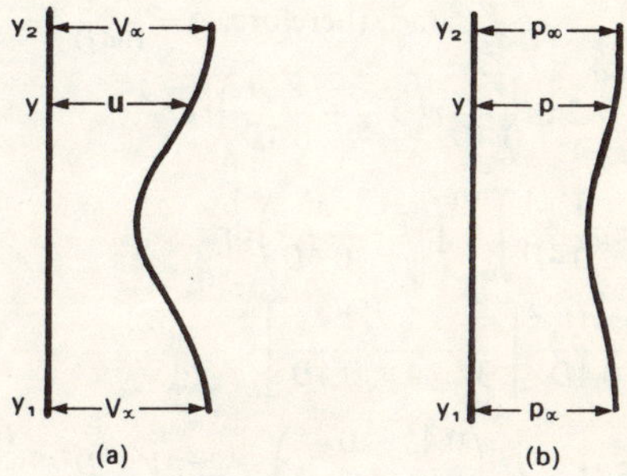

FIG. 76. (*a*) Velocity distribution across wake. (*b*) Pressure distribution across wake

With reference to Fig. 76 which illustrates the velocity distribution and the decrease in pressure which may occur in the wake, the forces acting are the skin friction, form drag (if any) and the force due to the imbalance of upstream and downstream pressures. If the total drag is F (= skin friction + form drag),

$$\int_{y_1}^{y_2} \rho u(V_\infty - u)\mathrm{d}y = F + \int_{y_1}^{y_2} (p_\infty - p)\mathrm{d}y$$

If the stream passes a body of revolution with its axis parallel to the direction of V_∞, the flow is three-dimensional but, otherwise, the procedure is the same, *see* Example 46.

EXAMPLE 46

A body of revolution is situated with its axis parallel to the direction of flow in a wind-tunnel test. In the wake, the velocity distribution is measured and is found to vary linearly from zero at the centre-line to a maximum value of V_∞ at a radius of 0·4D, where D is the diameter of the body. If there is no detectable pressure variation across the wake, show that the total drag force acting on the fluid is approximately $0{\cdot}084\rho V_\infty{}^2 D^2$.

SOLUTION

For two-dimensional flow it was shown above that the momentum equation may be written:

$$F = \int_{y_1}^{y_2} \rho u(V_\infty - u)\mathrm{d}y$$

For volumes of revolution the procedure is the same but the flux of momentum through an anular ring of radius r and width dr has to be considered and this leads to the equation:

$$F = \int_0^R \rho u(V_\infty - u)2\pi r\mathrm{d}r$$

where R is the radius of the wake at the section under consideration and u is the velocity in the wake at radius r. Applying this here:

$$\frac{u}{r} = \frac{V_\infty}{0{\cdot}4D} \text{ and, therefore, } u = \frac{V_\infty r}{0{\cdot}4D}$$

$$F = \int_0^{0{\cdot}4D} 2\pi\rho\frac{V_\infty}{0{\cdot}4D}r\left(V_\infty - \frac{V_\infty r}{0{\cdot}4D}\right)r\mathrm{d}r$$

$$= 2\pi\rho\frac{V_\infty^{\ 2}}{0{\cdot}4D}\int_0^{0{\cdot}4D}\left(r^2 - \frac{r^3}{0{\cdot}4D}\right)\mathrm{d}r$$

$$F = \frac{2\pi\rho V_\infty^{\ 2}}{0{\cdot}4D}\left[\frac{r^3}{3} - \frac{r^4}{4 \times 0{\cdot}4D}\right]_0^{0{\cdot}4D}$$

$$= 2\pi\rho V_\infty^{\ 2}D^2\left(\frac{0{\cdot}4^2}{3} - \frac{0{\cdot}4^2}{4}\right) = 2\pi\rho V_\infty^{\ 2}D^2 \times \frac{0{\cdot}4^2}{12}$$

whence $\underline{\underline{F = 0{\cdot}084\rho V_\infty^{\ 2}D^2}}$

In the above solution, had there been a pressure variation across the wake, the additional term would have been $\int_0^R (p_\infty - p)\,2\pi r\,\mathrm{d}r$.

BOUNDARY LAYER IN OPEN CHANNELS

For uniform flow in an open channel the slope i of the energy line is the same value as the slope s of the bed of the channel and

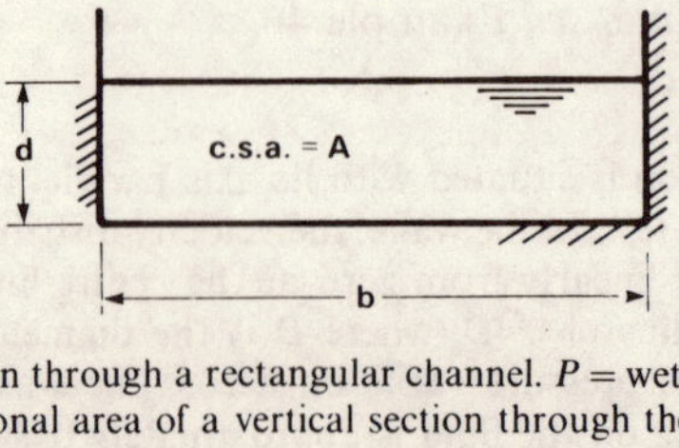

FIG. 77. Cross section through a rectangular channel. P = wetted perimeter = $b + 2d$. and A = cross-sectional area of a vertical section through the water

the depth of the liquid, d is constant along the length of the channel.

A is the area of a cross-section through the liquid, *see* Fig. 77. Considering unit length of the channel, the driving force causing flow is the component of the weight of the liquid down the channel, *i.e.* $\rho gAs \times 1 = \rho gAs$. A boundary layer forms on the sides and bottom of the channel and the resistance to flow is caused by the contact shear stress, τ_0, at the wetted boundary.

In unit length of channel, the shear force is $\tau_0 \times 1 \times P$, where P is the wetted perimeter.

For the flow to be uniform, the net force on the liquid must be zero.

i.e.
$$\rho gAs = \tau_0 P$$

Hence
$$\underline{\underline{\tau_0 = \rho gms}}$$

where m = hydraulic mean radius = A/P

In *wide* channels, it should be noted, the width is much greater than the depth and, therefore, the approximation can often be made that P = width and $m = A/P = d$.

EXAMPLE 47

The velocity, u in a uniform flow at height y above the bed of a wide channel can be represented by:

$$u = 5{\cdot}75\sqrt{\frac{\tau_0}{\rho}}\log_{10}\left(\frac{33y}{k}\right)$$

where k is the effective size of the roughness elements, τ_0 is the mean shear stress at the wetted boundary and ρ is the density of the liquid.

This distribution gives a velocity equal to the mean at $y = 0{\cdot}368\,d$, where d is the depth of flow.

(*a*) By putting $\tau_0 = m\rho gs$, where m is the hydraulic mean depth and s is the gradient of the bed, show that the Chézy coefficient in $m^{\frac{1}{2}}s^{-1}$ units is given by:

$$C = 18\log_{10}\frac{d}{k} + 19{\cdot}5$$

(*b*) A wide channel has a gradient of 1 in 2500. Measurements made in the channel, under what can be regarded as uniform flow conditions, gave the following results when the depth of flow was 1 m.

y (m)	0·03	0·11	0·40	0·75
u (m/s)	0·70	0·90	1·11	1·20

Show that the equation given for u is consistent with these observed values and determine the value of k in mm.

SOLUTION

(*a*) Since $\tau_0 = \rho gms$:

$$u = 5{\cdot}75\sqrt{gms}\,\log_{10}\frac{33y}{k}$$

since $u = \bar{u}$ when $y = 0{\cdot}368\,d$, the mean velocity, $\bar{u}$ is:

$$\bar{u} = 5{\cdot}75\sqrt{gms}\,\log_{10}\frac{33 \times 0{\cdot}368d}{k}$$

$$= 5{\cdot}75\sqrt{gms}\left\{\log_{10}\frac{d}{k} + \log_{10}(33 \times 0{\cdot}368)\right\}$$

Also, by the Chézy formula:

$$\bar{u} = C\sqrt{ms}$$

hence, $$C = 5{\cdot}75\sqrt{g}\left\{\log_{10}\frac{d}{k} + \log_{10}(33 \times 0{\cdot}368)\right\}$$

i.e. $$C = 18\log_{10}\frac{d}{k} + 19{\cdot}5$$

(*b*) In order to validate the formula, *u* is plotted against $\log_{10} y$, *see* Fig. 78.

y (m)	0·03	0·11	0·40	0·75
$\log y$	−1·523	−0·960	−0·400	−0·125
u (m/s)	0·70	0·90	1·11	1·20

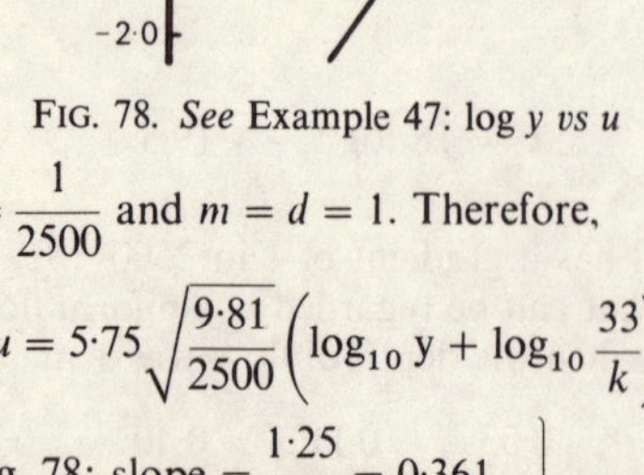

FIG. 78. *See* Example 47: log *y* *vs* *u*

In this example, $s = \dfrac{1}{2500}$ and $m = d = 1$. Therefore,

$$u = 5{\cdot}75\sqrt{\frac{9{\cdot}81}{2500}}\left(\log_{10} y + \log_{10}\frac{33}{k}\right)$$

From the graph, Fig. 78: slope $= \dfrac{1{\cdot}25}{3{\cdot}46} = 0{\cdot}361$

from the equation: slope $= 5{\cdot}75\sqrt{\dfrac{9{\cdot}81}{2500}} = 0.360$

} good agreement

When $\log_{10} y = 0$,

$$u = 1{\cdot}25 = 0{\cdot}361\,(\log_{10}33 - \log_{10}k)$$

$$\therefore \quad \log_{10}k = \log_{10}33 - \frac{1{\cdot}25}{0{\cdot}361} = 1{\cdot}13 \times 10^{-2}\ \text{m}$$

i.e. $\underline{\underline{k = 11\ \text{mm}}}$

QUESTIONS

1. Show that the total drag on a plane surface due to the presence of a boundary layer above the surface may be written as

$$F = \rho \int_0^{\delta} u(U - u)\,\mathrm{d}y$$

in which u is the velocity at a distance y from the surface, ρ is the fluid density, δ the layer thickness and U the free stream velocity. Under what circumstances is this expression useful?

What is meant by *displacement thickness*?

2. (*a*) With the aid of sketches of a flat plate in a uniform stream, define *boundary layer* and *wake*, *displacement thickness* δ^*, *momentum thickness* θ and *energy thickness* δ_E.

(*b*) What type of problems are solved with the aid of δ^*, θ and δ_E?

(*c*) The two-dimensional velocity distribution in the wake downstream of a flat plate at zero incidence in a fluid of density ρ may be approximated by

$$u = U\left\{1 - \cos(\tfrac{1}{2}\pi y/\delta)\right\}$$

The plate lies in the xz plane and U is the free-stream velocity in the x direction at distances $|y| \geqslant \delta$.

Calculate the total drag per unit width of plate and briefly indicate why the shear stress on the plate will not be constant. Assume the pressure in the free stream to be constant.

3. (*a*) Define boundary layer, displacement thickness δ^*, and momentum thickness θ.

(*b*) At constant pressure, a thin flat plate in the xz plane causes a boundary layer to form in a stream with uniform velocity U in the x direction. Show that the shear stress on the plate is

$$\tau_0 = \rho U^2(\mathrm{d}\theta/\mathrm{d}x)$$

(*c*) Find the thickness δ_1 of the boundary layer at a distance x_1 from the leading edge of the plate if the velocity distribution may be approximated by $u/U = (y/\delta)^{\frac{1}{7}}$ given that

$$\mathrm{d}\theta/\mathrm{d}x = (\rho U\delta/\mu)^{-\frac{1}{4}}/40$$

4. Assuming $F = \rho \int_{-\delta}^{\delta} u(U - u)\mathrm{d}y$ with the same notation as in Question 1 and $u/U = (y/\delta)^{\frac{1}{7}}$, derive an expression for F and find its value at a point 20 m downstream from the leading edge of a plate immersed in water flowing at a mean velocity of 4 m/s. Assume the kinematic viscosity of water is 1×10^{-6} m^2/s and $\dfrac{\delta}{X} = 0{\cdot}377(R_e)^{-\frac{1}{5}}$ where δ is the thickness of the layer at a distance X from the leading edge.

5. (*a*) A long body of depth D is moving through still air with velocity U. Its length is normal to the direction of motion and the relative flow may be considered two-dimensional and symmetrical about the body. Tests on a model indicate that the velocity distribution in the wake varies linearly from $U/2$ at the centre-line to U at a distance of D from the centre-line.

   Assuming the pressure in the wake is uniform, show that the drag force per unit length on the body is 2560 N when $D = 1$ m, $U = 80$ m/s and the density of the air is $1{\cdot}2$ kg/m^3.

   (*b*) If the pressure across the wake of the body in (*a*) varies linearly from a maximum value of p_0 in the uniform stream to a minimum at the centre-line of the wake, show that the total resistance per unit length of body is given by

$$F = \rho \int_{-\delta}^{\delta} u(U - u)\mathrm{d}y + \int_{-D}^{D} (p_0 - p)\mathrm{d}y$$

   where p is the pressure and u is the velocity at a distance y from the centre-line of the wake.

6. Derive an expression connecting the friction slope i in an open channel, the stress τ on the walls and bottom, the density ρ and the hydraulic mean depth m.

For a very wide channel it can be shown that the velocity distribution is given by

$$u = 5{\cdot}75 \sqrt{\frac{\tau}{\rho}} \log_{10}\left(\frac{33y}{K}\right)$$

where u is the velocity at height y above the bed which is covered with roughness of effective size K. If the mean speed of flow occurs at a height of 40 per cent of the stream depth, d, derive an expression for the coefficient C in Chézy's friction formula in the form

$$C = 18 \times \log_{10}\left(\frac{13{\cdot}2d}{K}\right)$$

7. Explain using clear sketches what is meant by the *displacement thickness* δ_1 and the *momentum thickness* δ_2 in relation to the boundary layer concept. Derive a general expression for δ_1 and then show that the assumption of a velocity distribution of the form

$$\frac{u}{U} = \left(\frac{y}{\delta}\right)^{\frac{1}{n}}$$

in the boundary layer leads to the expression

$$\delta_1 = \frac{\delta}{1 + n}$$

A simple broad crested weir 4 m long takes the form of a rectangular horizontal block the full width of a stream 4 m wide in which the depth of the flow upstream is 3 m, and the weir crest is 1 m above the bed. Assuming $Q = 1{\cdot}7\ BH^{\frac{3}{2}}$ where H is the total head upstream above the weir crest, make an approximate calculation of the thickness δ of the boundary layer at the downstream end of the crest given $\delta = 0{\cdot}37 \times (R_e)^{-\frac{1}{5}}$ and $\nu = 1{\cdot}14 \times 10^{-6}$ SI units.

[*Hint*: For a first approximation take H as upstream depth of 2 m above level of weir.]

8. (*a*) Describe *laminar* and *turbulent* boundary layers
   (*b*) A thin flat plate, 1 m square, is held in a stream of air (density, $\rho = 1{\cdot}2$ kg/m^3, kinematic viscosity, $\nu = 16$ mm^2/s) with a uniform velocity $U = 20$ m/s parallel to the plate.

   Assuming a negligibly short transition region, calculate the total drag on the plate given that:

   (*i*) the velocity u at a distance y from the plate, in the laminar boundary layer, is $u = U \sin(\pi y/2\delta)$ where δ is the boundary layer thickness;
   (*ii*) the momentum thickness in the turbulent boundary layer is $\delta_m = 0{\cdot}097\delta$;
   (*iii*) the thickness of the turbulent boundary layer at a distance x from the leading edge is $\delta = 0{\cdot}38x/R_e^{\frac{1}{5}}$ where the Reynolds number $R_e = Ux/\nu$; and
   (*iv*) the transition from laminar to turbulent boundary layer occurs when $R_e = 500\,000$.

CHAPTER 11

SOME APPLICATIONS OF BOUNDARY LAYER THEORY

IN a compact book such as this it is not possible to give a comprehensive account in one chapter of all the applications of boundary layer theory and, therefore, this chapter is confined to flow past spheres and cylinders and flow through pipes. These are topics which, because they are dealt with elsewhere in the book, merit further amplification.

FLOW ROUND A SPHERE

The reader will recall that the Reynold's number for flow round a sphere situated in a uniform stream of velocity V_∞ is $R_e = \dfrac{\rho V_\infty d}{\mu}$ where d is the diameter of the sphere, ρ is the density of the fluid and μ is the absolute viscosity of the fluid.

The streamlines for very slow flow ($R_e < 1$) past a sphere are indicated in Fig. 59 and it is only for this type of flow that Stokes' equation, $D = 3\pi\mu d V_\infty$, is valid. In this case the boundary layer totally encloses the sphere.

If the velocity is increased until R_e is significantly greater than 1, however, the flow pattern changes completely and will soon adopt one of the two patterns shown in Fig. 79 for flow past a sphere. If the sphere is smooth the flow will be as shown in Fig. 79 (*a*) with the boundary layer breaking away on the front half of the sphere. If, on the other hand, the surface of the sphere is rough and it is subjected to the same flow the boundary layer will be extended onto the latter half as shown in Fig. 79 (*b*).

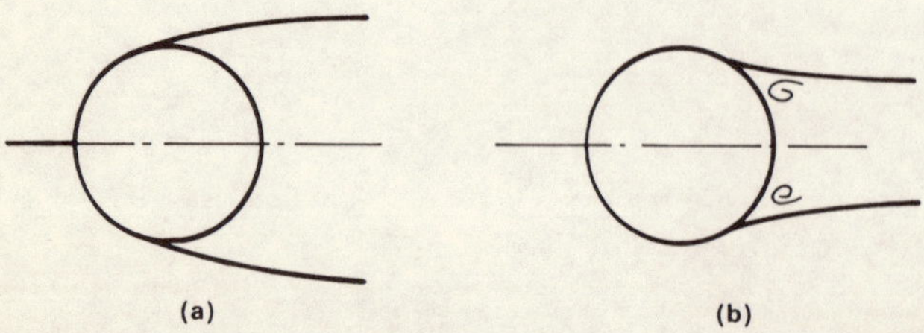

FIG. 79. Faster flow past a sphere: (*a*) "Smooth" sphere and (*b*) "Rough" sphere

The explanation of this phenomenon lies in the nature of the boundary layer which in Fig. 79 (*a*) is laminar and in (*b*), because of the roughened surface, is turbulent. The velocity distribution across the boundary layer will be similar to one of those shown in Fig. 73 (*b*) from which it can be seen that, for a specific value of the Reynolds number at which it is possible for the flow in the boundary layer to be either laminar or turbulent, the velocity at a given distance from the boundary will be greater in the turbulent case than in the laminar. The fluid has greater inertia due to the higher velocity and, therefore, does not break away from the boundary as early as the laminar flow. This is a fact of some importance in hydrodynamic and aerodynamic design and also in experiments with models in which surfaces are sometimes artificially roughened to reduce the drag force. Referring to Fig. 79, four definitions are given below.

Separation (or *breakaway*): the parting of the flow from the boundary.

Wake: the highly disturbed region, behind the sphere, which is contained between the limiting streamlines of the uniform flow. It should be noted that it would be wrong to describe the wake as turbulent because the word turbulence is reserved for the rapid, haphazard, tiny fluctuations of velocity which occur in the turbulent boundary layer, whereas the wake contains large eddies shed by the sphere from the points where breakaway occurs.

Form drag: Significant energy and momentum losses are associated with the large eddies in the wake and may result in a considerable resistance to the relative flow of the stream past the sphere. This resistance force is called the form drag. The total drag, *D*, sometimes called profile drag, is the sum of the skin friction and the form drag and, in good hydrodynamic/aerodynamic design, a

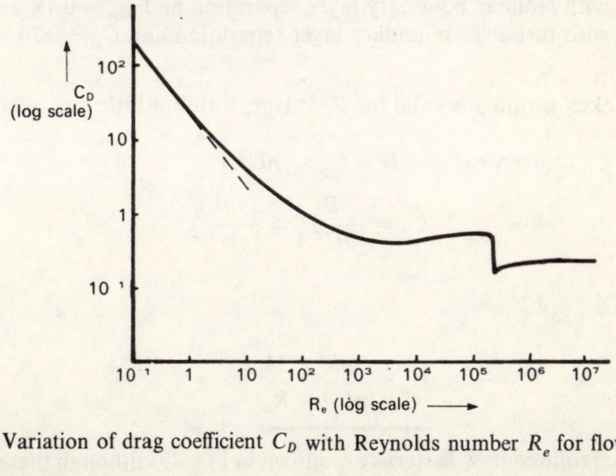

FIG. 80. Variation of drag coefficient C_D with Reynolds number R_e for flow past a sphere

balance is aimed at between these two components of the total drag.

Coefficient of drag, C_D: is defined by the equation

$$\text{drag, } D = C_d \times \tfrac{1}{2}\rho V_\infty{}^2 A$$

where ρ and V_∞ are as defined on p. 172 and A is the projected area of the body measured across the direction of the stream. Figure 80 shows how C_D varies with R_e for flow past a sphere.

Streamlined and bluff bodies

The occurrence of separation and wakes is usually associated with divergent flow. Abrupt divergence produces a wider wake (and greater energy loss) than a more gradual divergence and, for this reason, one way of reducing drag on an immersed body is to extend the tail so that the flow past it diverges slowly. Such a body is shown in Fig. 81 (*d*) and is called a streamlined body. By contrast, bodies which are not streamlined are often referred to as bluff bodies.

EXAMPLE 48

(*a*) Stokes derived the drag D experienced by a sphere of diameter d moving at a uniform velocity U through a fluid of viscosity μ to be $D = 3\pi\mu Ud$. State the most important condition of validity of this expression in relation to the particle Reynolds number. Calculate the coefficient of drag, C_D, from Stokes' formula.

(*b*) With the aid of sketches describe the flow past isolated spheres falling in a large volume of still fluid, and especially the wake, and explain the value of C_D:

(*i*) with laminar boundary layer separation and $C_D = 0{\cdot}45$, and
(*ii*) with turbulent boundary layer separation and $C_D = 0{\cdot}20$.

SOLUTION

(*a*) Stokes' formula is valid for $R_e < 1$ or, with but little loss of accuracy, $R_e < 2$.

Since C_D is given by: $\quad D = C_D \times \tfrac{1}{2}\rho U^2 A$

$$C_D = \frac{D}{\tfrac{1}{2}\rho U^2 A} = \frac{3\pi\mu Ud}{\tfrac{1}{2}\rho U^2 A}$$

Since $A = \dfrac{\pi}{4}d^2$,

$$C_D = \frac{24\mu}{\rho Ud} = \frac{24}{R_e}$$

(*b*) The required flow patterns are shown in Fig. 79, although these should be turned through 90° to indicate vertical flow, and the description is in the text on p. 173.

(*i*) Laminar boundary layer separation implies that the cylinder behaves as a smooth cylinder and this accounts for the relatively high value of C_D.

(*ii*) Turbulent boundary layer separation implies that the cylinder is behaving as a rough cylinder with separation occurring on the downstream half and, consequently, the lower value of C_D.

FLOW ROUND A CIRCULAR CYLINDER

A sphere and a cylinder both appear, in section, as circles and the flow pattern round them, on paper, might seem to be the same. They differ, however, because the distortions of the streamlines caused by introducing a sphere into a steady, uniform flow are three-dimensional, whereas the distortions to the flow past a cylinder (assuming it to be long) lie in two dimensions only.

Figure 81 (*a*), (*b*), (*c*) shows three stages in the development of a wake behind a circular cylinder and (*d*) shows the greatly reduced wake behind a streamlined body.

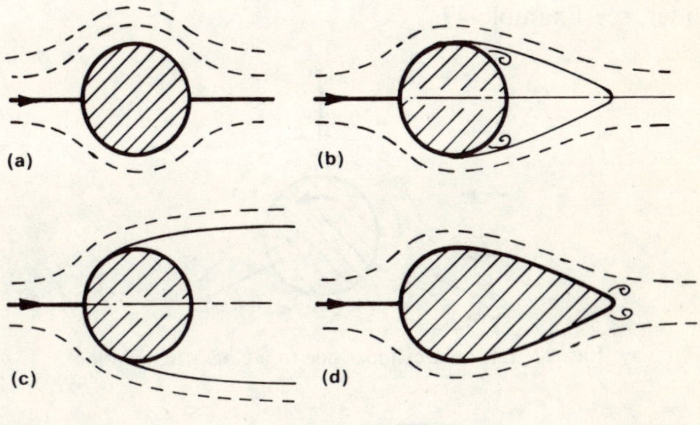

FIG. 81. (*a*), (*b*), (*c*): three stages in the development of a wake behind a circular cylinder. (*d*): reduced wake behind a "streamlined" body

Lift

Flow past a cylinder which is also rotating about its longitudinal axis is of fundamental importance in hydrodynamics and aerodynamics because the combination produces *lift*, the phenomenon which explains why aeroplanes fly and paddle steamers move through the water. The same phenomenon is also to be observed when rotation is added to the uniform motion of a sphere through a fluid, as is well-known to any ball-game player who uses spin, for example, to "flight" or "swerve" the ball. It is then known as the *magnus effect* and may be understood from a study of the streamlines.

The rotation of the cylinder is transmitted by viscous shear to the surrounding fluid which consequently adopts the flow pattern which would be given (outside the cylinder) by a two-dimensional free vortex situated at its centre. The streamlines are circles, concentric with a section through the cylinder, and indicate the presence of a circulation. If the streamlines for a uniform flow past a cylinder, shown in Fig. 59, are combined with these circular streamlines the resulting flow pattern is as shown in Fig. 82 from which flow is speeded up on one side (the top in Fig. 82) and slowed down on the other side of the cylinder. An imbalance results in the pressure above and below the cylinder and it is this which causes the lift.

It will be recalled, from p. 123 that $\Gamma = 2\pi\omega r^2 = 2\pi\mu$ and, therefore, that the circulation is independent of r. Moreover, the flow in a free vortex is irrotational, the uniform flow past a stationary cylinder is irrotational and, therefore, the combination of the two gives a potential flow. Consequently, the Bernoulli equation can be applied between streamlines and a mathematical expression can be deduced for the velocity of flow at any point on the surface of the cylinder, see Example 49.

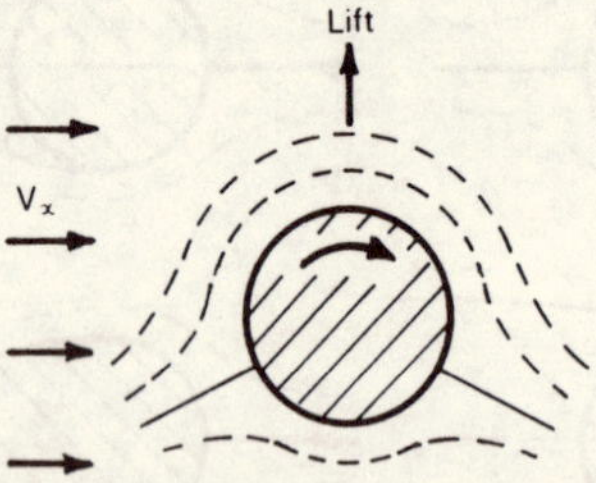

FIG. 82. Lift on a cylinder due to a clockwise rotation

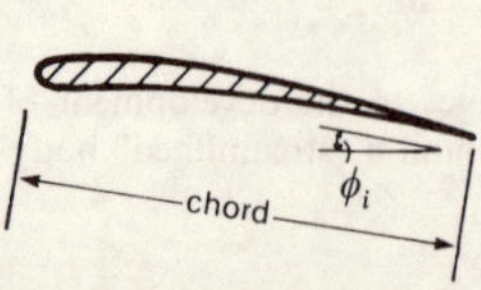

FIG. 83. Aerofoil section ϕ_i = *angle of incidence*, the *chord* is defined as shown and the *span* is measured at right angles to the section, *i.e.* into the paper

An aerofoil experiences the same lift effect as the rotating cylinder, the circulation in its case being induced by its shape. It arises from the initial mismatch at the trailing edge between the velocities of flow above and below the aerofoil.

Figure 83 indicates how, for an aerofoil, the *angle of incidence*, the *chord* and the *span* are measured. The projected area S of the aerofoil is defined as the product chord × span and a *coefficient of lift* C_L is defined as

$$C_L = \frac{L}{\frac{1}{2}\rho V_\infty{}^2 S}$$

where L is the lift force and V_∞ is the flow velocity of the uniform stream remote from the cylinder.

EXAMPLE 49

(*a*) A circular cylinder of radius r rotates at an angular velocity ω about its horizontal axis in a uniform stream of water flowing at an undisturbed velocity U. Derive an expression for the distribution of pressure around its circumference and hence calculate the lift per unit length of cylinder and show that the form drag is zero.

(*b*) Would there be any form drag if the cylinder had a non-circular cross-section?

SOLUTION

(*a*) It is shown on p. 120 that the velocity at a point Q, say, on the surface of the cylinder due to the uniform stream U is $2U \sin\theta$ where θ is measured from a reference line parallel to the flow. The velocity component at Q due to a to a circulation Γ is $\dfrac{\Gamma}{2\pi r}$.

Hence the actual velocity at Q is:

$$v = 2U\sin\theta + \frac{\Gamma}{2\pi r}$$

If the upstream pressure is p_0 and the pressure at Q is p the Bernoulli equation gives:

$$p_0 + \tfrac{1}{2}\rho U^2 = p + \tfrac{1}{2}\rho v^2$$

Hence:
$$p = p_0 + \tfrac{1}{2}\rho\left\{U^2 - \left(2U\sin\theta + \frac{\Gamma}{2\pi r}\right)^2\right\}$$

Lift force is:

$$L = -\int_0^{2\pi} p\sin\theta . r\,d\theta \qquad (1)$$

$$= -\int_0^{2\pi}\left\{p_0 + \tfrac{1}{2}\rho\left(U^2 - 4U^2\sin^2\theta - \frac{4U\Gamma\sin\theta}{2\pi r} - \frac{\Gamma^2}{4\pi^2 r^2}\right)\right\}\sin\theta r d\theta$$

Since $\int_0^{2\pi} \sin\theta\, d\theta = 0$, $\int_0^{2\pi} \sin^3\theta\, d\theta = 0$ and $\int_0^{2\pi} \sin\theta^2 d\theta = \pi$

$$L = \tfrac{1}{2}\rho\int_0^{2\pi}\frac{4U\Gamma\sin^2\theta}{2\pi}d\theta = \rho U\Gamma$$

The equation $L = \rho U\Gamma$ is known as the Kutta Joukowski equation.

To be consistent with the sign convention used for θ, Γ is positive for anti-clockwise circulation. Positive uniform flow is from left to right and these two combine to give a downward "lift" force—hence the minus sign in equation (1).

Similarly, the form drag is:

$$D = \tfrac{1}{2}\rho \int_0^{2\pi} p \cos\theta \, r\mathrm{d}\theta = 0$$

because $\displaystyle\int_0^{2\pi} \cos\theta \, \mathrm{d}\theta = 0$

$$\int_0^{2\pi} \sin\theta\cos\theta \, \mathrm{d}\theta = \int_0^{0} \sin\theta \, \mathrm{d}(\sin\theta) = 0$$

and $\displaystyle\int_0^{2\pi} \sin^2\theta\cos\theta \, \mathrm{d}\theta = \int_0^{0} \sin^2\theta \, \mathrm{d}(\sin\theta) = 0$

(*b*) No. The above equation applies only if the fluid is ideal. For real fluids the Kutta Joukowski equation has to be modified to allow for the formation of a boundary layer, breakaway, wake, *etc*.

The theory is applicable, with modifications, to aerofoil sections. At low angles of incidence the form drag is small compared with the lift. As the angle of incidence is increased, the drag increases and so does the discrepancy between the lift as measured by experiment and as calculated by theory.

PIPE FLOW

In regard to what might be called the *modern* pipe flow formulae the mathematical intricacies are made more complex by their heavy dependence on the results of experiments for the evaluation of the various constants in them. The mixture of empirical results and mathematical reasoning can be, at first reading, a little overwhelming and for this reason a step-by-step approach is used below. Each step is numbered.

1. Where a fluid enters a round pipe a boundary layer starts to form as described on p. 152 and grows downstream of the pipe entrance until it reaches a thickness equal to the radius of the pipe, *R*. If the flow is slow enough the boundary layer will be laminar not only throughout its developing stages but, also, beyond the point where its thickness becomes *R*; *i.e.* the flow is laminar throughout the whole pipe and, therefore, the velocity distribution will be parabolic, Poiseuille's equation, *see* p. 32 will apply and the shear stress will be $\tau = \mu \dfrac{\mathrm{d}u}{\mathrm{d}y}$ where μ is the dynamic viscosity and $\dfrac{\mathrm{d}u}{\mathrm{d}y}$ is the velocity gradient.

2. For pipe-flow, Reynolds number is defined by $R_e = \frac{\rho U d}{\mu}$ where ρ and μ are the density and the viscosity of the fluid, respectively; U is the mean velocity of flow and d is the diameter of the pipe. The lower critical value of the Reynolds number lies between about 2000 and 2500 and this means that if $R_e < 2000$ the flow will be laminar. Laminar flow can also occur in a smooth pipe at very much higher values of R_e and has been obtained, for example, with $R_e = 40\,000$. At higher values of R_e, however, it is normal for the flow in round pipes to be turbulent and the turbulence may usually be considered to occupy the whole length, any laminar boundary layer which might form at entry to the pipe normally being of such a short length as to be negligible.

3. In the theory of pipe-flow it is common practice to assume that the pipe lies in the x-direction. The other co-ordinate axes, y and z are at right-angles to the direction of flow and, for round pipes, because of the radial symmetry, it is usually sufficient to work in terms of only one of these, normally the y-axis, y being measured from the wall. *i.e.*, at the wall, $y = 0$. Again, it is common practice in two- or three-dimensional flow considerations to adopt U and V as the symbols for the mean velocity components in the x and y directions respectively, u and v as the velocity components at a point and in pipe-flow theory, therefore, U is the fluid mean velocity and $V = 0$.

4. To understand the theory of turbulent pipe-flow it is necessary to understand the mechanics of the turbulent motion of the fluid particles. Whereas the flow velocity at a point may be averaged over a period of time and called u (which is sometimes referred to, therefore as the *temporal* mean), tiny fluid masses do not move all the time with the velocity u but, in fact, have velocity components u' and v' superimposed on u. u' and v' are the fluctuating velocity components associated with turbulence. If a control volume is established within the flow pattern of an incompressible fluid it is clear that for continuity the net flows in and out must be the same, *i.e.* $\Sigma u' = 0$ and $\Sigma v' = 0$. Nevertheless, on adjoining layers the velocities in the x-direction can be $u + u'$ and $u - u'$ respectively and, therefore, momentum can be transferred between the layers. The momentum transferred across a small area A will be $Au'v'$ and since *force = flux of momentum*, shear force $= \rho A u'v'$ and the shear stress $\tau = \rho u'v'$.

Prandtl found it convenient to introduce, as a measure of the region involved in the turbulent exchange, a characteristic length which he named the *mixing length* l and which can be defined as the average length over which the fluid masses carry their momentum when they are displaced perpendicularly to the main flow. It is fairly easy to establish that $u' = l\frac{du}{dy}$ and $v' = l\frac{dv}{dy}$ and, therefore,

$$\tau = \rho u'v' = \rho l^2\left(\frac{du}{dy}\right)^2$$

5. The *eddy viscosity* η has been referred to on p. 152 and is defined by $\tau = \eta\frac{du}{dy}$. It is a concept which was introduced into the consideration of turbulent flow to make it analogous to viscous flow. Now, as a consequence of the previous paragraph, it can be expressed in terms of the mixing length thus:

$$\eta = \rho l^2 \frac{du}{dy}$$

6. If a laminar sublayer exists then, in it, the shear stress becomes $\tau_0 = \mu\frac{du}{dy}$. If, for simplicity, a linear relationship is assumed between u and y such that $\frac{du}{dy} = \frac{u}{y}$,

$$\tau_0 = \mu\frac{u}{y}, \quad \text{or} \quad \frac{\tau_0}{\rho} = \nu\frac{u}{y}$$

The quantity $\sqrt{\frac{\tau_0}{\rho}}$ has the dimensions of velocity and, therefore, is called the *shear velocity* or *shear stress velocity* or, often, the *friction velocity*, u_*. If the thickness of the laminar sublayer is δ, then for $y \leqslant \delta$:

$$u_*^2 = \nu\frac{u}{y}$$

i.e.

$$\frac{u}{u_*} = \frac{u_* y}{\nu}$$

Outside the boundary sublayer it is observed that $l \propto y$,

since

$$u_*^2 = \frac{\tau_0}{\rho} = l^2\left(\frac{du}{dy}\right)^2$$

$$\frac{du}{dy} = \frac{u_*}{l} \propto \frac{u_*}{y}$$

i.e.

$$\frac{du}{u_*} = \text{constant} \times \frac{dy}{y}$$

$\therefore$ for $y \geqslant \delta$ $$\frac{u}{u_*} = A_1 \log_e y + B_1$$

where A_1 and B_1 are constants.

7. Summarising the previous section:

for $y \leqslant \delta$ $$\frac{u}{u_*} = \frac{u_* y}{\nu}$$

for $y \geqslant \delta$ $$\frac{u}{u_*} = A_1 \ln y + B_1$$

When $y = \delta$ there will be a particular value of u which satisfies both these equations. This particular value is sometimes called the *wall velocity* u_w and to simplify the algebra let $\frac{u_w}{u_*} = N$. Then when $y = \delta$

$$N = \frac{u_* \delta}{\nu}$$

and $$N = A_1 \ln \delta + B_1 = A_1 \ln \left(\frac{N\nu}{u_*}\right) + B_1$$

$\therefore$ $$B_1 = N - A_1 \ln \left(\frac{N\nu}{u_*}\right)$$

hence $$\frac{u}{u_*} = A_1 \ln y + N - A_1 \ln \left(\frac{N\nu}{u_*}\right)$$

$$= A_1 \ln \left(\frac{yu_*}{\nu}\right) + B_2$$

The constants A_1 and B_2 have been evaluated by experiment to give the Karman Prandtl equation for turbulent flow near smooth boundaries, *i.e.*

$$\frac{u}{u_*} = 5{\cdot}75 \log_{10} \left(\frac{yu_*}{\nu}\right) + 5{\cdot}5$$

8. In rough pipes the size k of the roughness projections may be expressed as a function of y such that $y = km$ when $u = u_w$.

From 6: $$\frac{u}{u_*} = A_1 \ln y + B_1$$

and, if the maximum value of u is denoted u_m, $u = u_m$ when $y = R$

$$\frac{u_m}{u_*} = A_1 \ln R + B_1$$

Subtracting, $$\frac{u_m - u}{u_*} = A_1 \ln (R/y)$$

It should be noted that $(u_m - u/u_*$ is sometimes called the *velocity deficiency*.

Since $y = km$, $$\frac{u_m - u_w}{u_*} = A_1 \ln \left(\frac{R}{mk}\right)$$

$$\frac{u_m}{u_*} = \frac{u_w}{u_*} + A_1 \ln \left(\frac{R}{mk}\right)$$

Also $$\frac{u_m}{u_*} = \frac{u}{u_*} + A_1 \ln \left(\frac{R}{y}\right)$$

subtracting, $$\frac{u}{u_*} = \frac{u_w}{u_*} + A_1 \ln \left(\frac{y}{mk}\right) = A_1 \ln \left(\frac{y}{k}\right) + B_3$$

where $B_3 = \frac{u_w}{u_*} - A_1 \ln m$

The constants A_1 and B_3 have been evaluated by experiment to give the Karman Prandtl equation for turbulent flow in rough pipes, namely:

$$\frac{u}{u_*} = 5{\cdot}75 \log_{10}\left(\frac{y}{k}\right) + 8{\cdot}50$$

9. It is interesting to note that, if expressions are derived for u in terms of the mean velocity U, a result which is applicable to both smooth and rough pipes is obtained, the equation being:

$$\frac{u - U}{u_*} = 5{\cdot}75 \log_{10}\frac{y}{k} + 3{\cdot}75$$

10. *Karman Prandtl resistance equations.* The foregoing theory indicates the extent to which experimental results have to be used in order to understand and describe flow in pipes. Two other important results, obtainable by the same sort of combination of mathematical processes and experimental data, are the equations which give the friction factor (from the Darcy–Weisbach equation) as a function of the Reynolds number, R_e and the roughness k of a pipe of radius R. These are sometimes called the Karman Prandtl resistance equations and are:

for smooth pipes $$\frac{1}{\sqrt{f}} = 4 \log_{10} (2R_e\sqrt{f}) - 1{\cdot}6$$

for rough pipes

$$\frac{1}{\sqrt{f}} = 4\log_{10}\left(\frac{R}{k}\right) + 3{\cdot}48$$

where f is the Darcy–Weisbach f given by $h_f = \dfrac{4flu^2}{2gd}$.

EXAMPLE 50

Two empirical formulae are in common use for the estimation of the hydraulic gradient $i = h/l$ in the steady flow through a straight pipe of diameter d. They are the Chézy formula, $V = C(mi)^{\frac{1}{2}}$, and the formula due to Manning and Strickler, $V = n^{-1}m^{\frac{2}{3}}i^{\frac{1}{2}}$, where:

V is the mean velocity;
$m = d/4$, the hydraulic radius or mean depth;
h is the head lost in length l; and
C, n are the respective friction coefficients.

The Darcy–Weisbach formula, $h = 4flV^2/(2gd)$, has a dimensionless friction coefficient f which may be expressed in any units by:

$$1/\sqrt{f} = 4\log_{10}(2R_e\sqrt{f}) - 1{\cdot}6$$

for smooth pipe flow, where $R_e = \dfrac{Vd}{\nu}$; *and*

$$1/\sqrt{f} = 4\log_{10}(r/k) + 3{\cdot}48$$

for rough pipe flcw, where k is typical of the wall roughness size.

(*a*) Establish and discuss the equations linking C, n and f with each other and with R_e, $r = d/2$ and k.

(*b*) Show that, in metric units, $1/n = d^{-\frac{1}{6}} \times 22{\cdot}3\log(3{\cdot}7d/k)$

SOLUTION

The Darcy–Weisbach formula can be written:

$$V^2 = \frac{2g}{f}.\frac{d}{4}.\frac{h}{l}$$

$$\therefore \qquad V = \sqrt{\frac{2g}{f}}\, m^{\frac{1}{2}}i^{\frac{1}{2}}$$

Compare with

$$V = C\sqrt{mi}$$

$$\therefore \qquad \underline{C = \sqrt{\frac{2g}{f}}}$$

Since $1/\sqrt{f}$ can be expressed in terms of R_ϵ or r/k, clearly, so can C. It should be noted, however, that since f is dimensionless, C has the dimensions of $\sqrt{2g}$ and, therefore, has a value which depends not only on the roughness of the boundaries or the flow conditions but, also, on the units used. In metric units, for example, $C = 4{\cdot}43/\sqrt{f}$ whereas, in the old imperial units, $C = 8/\sqrt{f}$.

The independent discovery of the other formula is attributed to Manning in this country but, abroad, is often attributed to Strickler—hence the double name. At the time of discovery it was thought that n depended only on wall roughness and, although subsequent research has shown this not to be strictly true, it is sufficiently accurate in the middle ranges of R_e for most practical purposes. The advantage of using $1/n$ for the coefficient is that n increases as roughness increases.

Despite the independent discovery of this formula it can be related to Chézy's formula by putting $C = \frac{1}{n}m^{\frac{1}{6}}$.

Then,
$$V = \frac{1}{n}m^{\frac{1}{6}}m^{\frac{1}{2}}i^{\frac{1}{2}} = \frac{1}{n}m^{\frac{2}{3}}i^{\frac{1}{2}}$$

Also:
$$\frac{1}{n} = Cm^{-\frac{1}{6}} = \left(\frac{d}{4}\right)^{-\frac{1}{6}}\sqrt{\frac{2g}{f}}$$
$$= d^{-\frac{1}{6}} \times 5{\cdot}58\left\{4\log\left(\frac{r}{k}\right) + 3{\cdot}48\right\}$$
$$= d^{-\frac{1}{6}} \times 22{\cdot}3\left\{\log\left(\frac{r}{k}\right) + \frac{19{\cdot}4}{22{\cdot}3}\right\}$$
$$= d^{-\frac{1}{6}} \times 22{\cdot}3\left\{\log\left(\frac{d}{2k}\right) + \log 7{\cdot}4\right\}$$
$$= 22{\cdot}3d^{-\frac{1}{6}} \times \log\left(\frac{3{\cdot}7d}{k}\right)$$

Again it should be appreciated that $1/n$ is not dimensionless and the numerical coefficients in the above equation depend on the choice of units; metric units are used here.

EXAMPLE 51

For steady fully-developed turbulent flow in a pipe of circular cross-section the velocity distribution can be represented by:
$$\frac{v}{v_{max}} = \left[1 - \frac{r}{R}\right]^{\frac{1}{n}}$$

where v is the velocity at radius r and n has a value between 6 and 9. depending on the flow conditions.

(*a*) Show that the mean velocity $\bar{v}$ is given by:
$$\bar{v} = \frac{2n^2}{(n+1)(2n+1)} \times v_{max}$$

(*b*) Derive an expression, in terms of R and n for the radius r_* at which the velocity is equal to the mean velocity.

(*c*) It is proposed to use a "point-velocity-measuring instrument" (*e.g.* a pitot-static tube) to obtain a value for the mean velocity and, hence, the flow rate of a steady flow of a fluid in a pipe. The instrument is to be positioned at a radius equal to 0·75 R and gives the velocity $v_{0\cdot75}$.

If it is assumed that the above formulae apply and that the value of n, although unknown, is likely to lie between 6 and 9, show that $v_{0\cdot75}$ will be within 1 per cent of $\bar{v}$.

SOLUTION

(*a*) Mean velocity is given by: $\bar{v} \times \pi R^2 = \int_0^R 2\pi r v \,\mathrm{d}r$

Since $\dfrac{v}{v_{max}} = \left(1 - \dfrac{r}{R}\right)^{1/n}$,

$$\frac{\bar{v}}{2v_{max}} = \frac{1}{\pi R^2}\int_0^R \pi r\left(1 - \frac{r}{R}\right)^{1/n} \mathrm{d}r$$

Let $x = \dfrac{r}{R}$ so that $\mathrm{d}r = R\,\mathrm{d}x$ and limits become 0 to 1.

Hence
$$\frac{\bar{v}}{2v_{max}} = \frac{1}{R^2}\int_0^1 R^2 x(1-x)^{1/n}\mathrm{d}x$$
$$= \int_0^1 x(1-x)^{1/n}\mathrm{d}x$$

Integrating "by parts":
$$\frac{\bar{v}}{2v_{max}} = \left[-\frac{n}{(n+1)}x(1-x)^{\frac{1+n}{n}}\right]_0^1 + \frac{n}{n+1}\int_0^1 (1-x)^{\frac{1+n}{n}}\mathrm{d}x$$
$$= 0 - \frac{n}{n+1}\,\frac{n}{2n+1}\left[(1-x)^{\frac{1+2n}{n}}\right]_0^1$$
$$= \frac{n^2}{(n+1)(2n+1)}$$

Hence:
$$\bar{v} = \frac{2n^2}{(n+1)(2n+1)} \times v_{max}$$

(*b*) When $v = \bar{v}$,
$$\frac{\bar{v}}{v_{max}} = \left(1 - \frac{r_*}{R}\right)^{1/n}$$

$$\left(1 - \frac{r_*}{R}\right)^{1/n} = \frac{2n^2}{(n+1)(2n+1)}$$

(c) $$\frac{v_{0\cdot75}}{v_{max}} \div \frac{\bar{v}}{v_{max}} = \left(1 - \frac{0\cdot75R}{R}\right)^{1/n} \div \frac{2n^2}{(n+1)(2n+1)}$$

$$= 0\cdot25^{1/n} \times \frac{(n+1)(2n+1)}{2n^2}$$

i.e. $$\frac{v_{0\cdot75}}{\bar{v}} = 0\cdot25^{1/n} \times \left(1 + \frac{3n+1}{2n^2}\right)$$

Tabulating:

n:	6	7	8	9
$\frac{v_{0\cdot75}}{\bar{v}}$:	1·0031	1·0044	1·0051	1·0054

which is within 1 per cent throughout the range.

11. *Commercial pipes.* A result, which is derived fairly easily from a combination of the Darcy–Weisbach equation and Poiseuille's equation, and which is left as an exercise for the reader is that, for laminar flow in round pipes, $f = \frac{16}{R_e}$.

In turbulent flow the Karman Prandtl resistance equations, *see* p. 182, indicate that the relationship between R_e and f can be represented by a single curve for hydraulically smooth pipes whereas, for hydraulically rough pipes, it assumes the form of a family of curves, each curve corresponding to a different value of $\frac{k}{d}$. Nikuradse verified that this was the case in his well known experiments on pipes artificially roughened with sand grains and the results in the form of a graph can be seen in many standard textbooks on fluid mechanics.

The Nikuradse graph is not applicable to commercial pipes, however, as they do not behave in exactly the same way as the artificially roughened pipes, the difference usually being attributed to the variable degree of roughness to be found in commercial pipes. The larger internal protrusions shed eddies at lower Reynolds numbers than the smaller protrusions and, consequently, the changeover from "smooth" to "rough" is less sharply defined than it is for pipes roughened with sand grains of a uniform size.

Further experimental work has been carried out on commercial pipes and has led to an empirical formula, usually attributed to Colebrook and White and named after them, *viz.*:

$$\frac{1}{\sqrt{f}} = -4\log_{10}\left(\frac{k/d}{3\cdot7} + \frac{1\cdot255}{R_e\sqrt{f}}\right)$$

The plot of this relationship on log–log graph paper exhibits the three elements referred to previously, *i.e.* a straight section corresponding to $f = \frac{16}{R_e}$ for laminar flow, a single curve for smooth pipes and a family of curves for rough pipes. The construction was first done by Moody and the diagram is named after him (*see* Fig. 84).

EXAMPLE 52

Given the Colebrook–White formula

$$\frac{1}{\sqrt{f}} = -4\log_{10}\left(\frac{k/d}{3{\cdot}7} + \frac{1{\cdot}255}{R_e\sqrt{f}}\right)$$

plot curves to show the variation of f with R_e for the following values of $\frac{k}{d}$: 10^{-6}, 10^{-4}, 10^{-3}, 10^{-2}.

SOLUTION

Since $$\frac{1}{\sqrt{f}} = -4\log_{10}\left(\frac{k/d}{3{\cdot}7} + \frac{1{\cdot}255}{R_e\sqrt{f}}\right)$$

$$\frac{k/d}{3{\cdot}7} + \frac{1{\cdot}255}{R_e\sqrt{f}} = 10^{-1/(4\sqrt{f})}$$

$$\therefore \quad \frac{1{\cdot}255}{R_e\sqrt{f}} = 10^{-1/(4\sqrt{f})} - \frac{k/d}{3{\cdot}7}$$

i.e. $$R_e = \frac{1{\cdot}255}{\sqrt{f}} \div \left(10^{-1/(4\sqrt{f})} - \frac{k/d}{3{\cdot}7}\right)$$

and, after ascribing a value to $\frac{k}{d}$, R_e can be calculated for each chosen value of f (and $\sqrt{f}$). For this kind of computation, a programmable calculator is invaluable and the above equation for R_e has been arranged in a form suitable for programming. The computation is given in Table IV.

Before plotting R_e *vs* f it is useful to calculate the asymptotic value of each branch of the curve. This can be done by letting $R_e \to \infty$.

Then: $\frac{1}{\sqrt{f}} = -4\log_{10}\left(\frac{k/d}{3{\cdot}7}\right)$, from which

$\frac{k}{d}$:	10^{-6}	10^{-4}	10^{-3}	10^{-2}
f:	0·001 45	0·003 00	0·004 91	0·009 48

f	$\sqrt{f}$	R_e			
		$\frac{k}{d}$: 10^{-6}	10^{-4}	10^{-3}	10^{-2}
0·0015	0·039	$4{\cdot}03 \times 10^8$			
0·0016	0·040	$1{\cdot}07 \times 10^8$			
0·0018	0·042	$2{\cdot}93 \times 10^7$			
0·0020	0·045	$1{\cdot}00 \times 10^7$			
0·0025	0·050	$2{\cdot}51 \times 10^6$			
0·0035	0·059		$6{\cdot}54 \times 10^5$		
0·004	0·063	$1{\cdot}85 \times 10^5$	$2{\cdot}47 \times 10^5$		
0·005	0·071	$5{\cdot}87 \times 10^4$	$6{\cdot}45 \times 10^4$	$5{\cdot}71 \times 10^5$	
0·006	0·077			$5{\cdot}03 \times 10^4$	
0·0075	0·087	$1{\cdot}08 \times 10^4$	$1{\cdot}10 \times 10^4$	$1{\cdot}35 \times 10^4$	
0·010	0·100	$3{\cdot}97 \times 10^3$	$4{\cdot}00 \times 10^3$	$4{\cdot}34 \times 10^3$	$2{\cdot}73 \times 10^4$
0·011	0·105	$2{\cdot}87 \times 10^3$	$1{\cdot}89 \times 10^3$	$3{\cdot}07 \times 10^3$	$8{\cdot}20 \times 10^3$
0·012	0·111			$2{\cdot}25 \times 10^3$	$4{\cdot}33 \times 10^3$
0·013	0·114				$2{\cdot}97 \times 10^3$

TABLE IV

The results in Table IV should be plotted on logarithmic graph paper to satisfy that the curves obtained do accord with the more elaborate plot of the Moody diagram in Fig. 84. The use of the Moody diagram is illustrated by Example 53.

EXAMPLE 53

An outlet pipe from a reservoir discharges to atmosphere as shown in Fig. 85. The 150 mm diameter concrete pipe is required to discharge 0·05 m^3/s. With reference to the Moody diagram, *see* Fig. 84, determine the maximum possible length of pipe, assuming an energy loss at the entry to the pipe of one half of the kinetic energy of the fluid. The kinematic viscosity of the water may be assumed as 1·14 mm^2/s. What flow conditions exist in the pipe?

SOLUTION

Average velocity of flow is given by:

$$V = Q/(\tfrac{1}{4}\pi d^2) = \frac{0{\cdot}05}{\tfrac{1}{4}\pi \times 0{\cdot}15^2} = 2{\cdot}83 \text{ m/s}$$

$$R_e = \frac{Vd}{\nu} = \frac{2{\cdot}83 \times 0{\cdot}15}{1{\cdot}14 \times 10^{-6}} = 3{\cdot}72 \times 10^5$$

and the flow is, therefore, turbulent.

Since concrete pipes are factory-made, take $k = 0{\cdot}3$ mm

$$\frac{k}{d} = \frac{0{\cdot}3}{150} = 0{\cdot}002$$

From the Moody diagram, *see* Fig. 84, these values of R_e and $\frac{k}{d}$ give $f = 0{\cdot}006$.

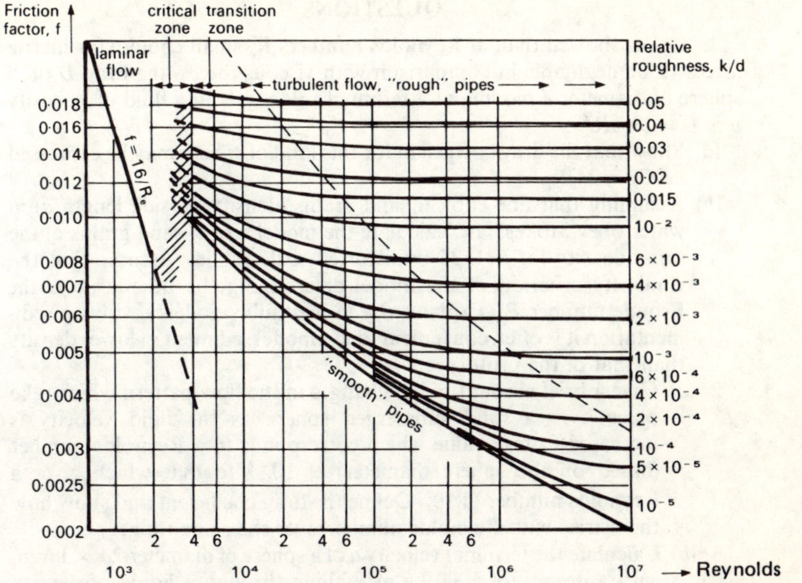

FIG. 84. Moody diagram

Typical values of roughness k (in mm) are:

concrete:	0·3–3	asphalted cast iron	0·1
cast iron	0·3	wrought iron	0·05
galvanised steel	0·2		

Since head available = total losses,

$$H = 26{\cdot}8 - 16{\cdot}7 = \tfrac{1}{2}\frac{V^2}{2g} + \frac{V^2}{2g} + \frac{4flV^2}{2gd}$$

$$\frac{4 \times 0{\cdot}006 \times l \times 2{\cdot}83^2}{2g \times 0{\cdot}15} = 10{\cdot}1 - \frac{3}{2} \times \frac{2{\cdot}83^2}{2g}$$

from which, $l = 145{\cdot}3\ m$, say 145 m

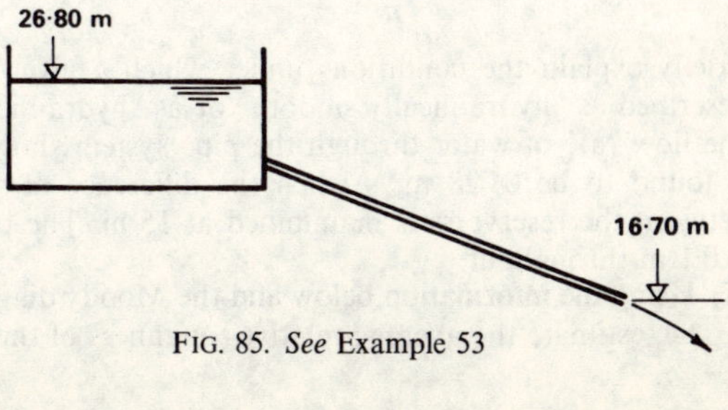

FIG. 85. *See* Example 53

QUESTIONS

1. Stokes showed that, at Reynolds numbers R_e small enough for inertia forces to be negligible in comparison with viscous forces, the drag D on a sphere of diameter d moving at a terminal velocity U in a fluid of viscosity μ is $D = 3\pi\mu dU$.
   (*a*) Show that the drag coefficient for this kind of sphere may be expressed as $C_D = 24/R_e$.
   (*b*) Assuming that fine cubical sand grains (density ρ_s, side length a) in water obey Stokes' law, calculate the model size of sand grains of the same density for a 1:25 undistorted scale model. Assume that the similarity criterion of the model determining the time scale is the Froude number F_r. Explain why in hydraulic models involving sedimentation it is often convenient to use model sediment of lower density than that of the prototype.
2. (*a*) Give a brief account of the changes in the flow pattern which take place over a solid submerged sphere as the fluid velocity is increased from a value which corresponds to a Reynolds number (based on the sphere diameter) of 10^{-2} to that which gives a Reynolds number of 10^6. Define the drag coefficient and show how this varies with Reynolds number over the same range.
   (*b*) Calculate the terminal velocity u of a sphere of diameter $2a$ = 1 mm, with a density of 6000 kg/m^3, falling through a liquid of density ρ = 1000 kg/m^3 and viscosity μ of 0·1 Ns/m^2. Assume that the flow would be in the Stokes law region and then show this to be true. The drag on a solid sphere in the Stokes law region is given be $D = 6\pi a\mu u$.
3. In the analysis of steady flow through cylindrical pipes, two formulae are in common use:
   (*a*) $V^2 = 2gmi/f$ (Darcy–Weisbach); and
   (*b*) $V = m^{\frac{2}{3}}i^{\frac{1}{2}}/n$ (Manning–Strickler), where i is the total-head gradient and the other symbols are defined below.
      (*i*) Briefly explain the origins and the advantage of these formulae.
      (*ii*) Establish the equation linking n with f, g and m and discuss this equation.
      (*iii*) Express the roughness coefficient n in terms of the mean velocity V, the pipe diameter d or hydraulic radius m, the pipe roughness k, and the pipe Reynolds number R_e and hence show that:

$$\frac{1}{n} = 22{\cdot}4\mathrm{d}^{-\frac{1}{6}}\log\frac{3{\cdot}7d}{k}$$

4. (*a*) Briefly explain the conditions under which a given pipe can be described as "hydraulically smooth" or as "hydraulically rough".
   (*b*) The flow rate of water through the pipe system shown in Fig. 86 is found to be 0·028 m^3/s when the difference of surface levels between the reservoirs is maintained at 15 m. The pipe diameter is 0·1 m throughout.
      (*i*) Using the information below and the Moody diagram, *see* Fig. 84, estimate the average relative toughness of the pipe.

(*ii*) If the relative roughness increases by 25 per cent what will be the likely percentage change in flow rate?

Assume that:

loss of head at entrance to pipe = $0{\cdot}5V^2/(2g)$;
loss of head at 45° bend = $0{\cdot}4V^2/(2g)$;
loss of head at a fully-open globe valve = $10V^2/(2g)$;
loss of head at exit from pipe = $V^2/(2g)$;
where V is the mean velocity of flow in the pipe.

Take the absolute viscosity of water as 0·001 kg/ms.

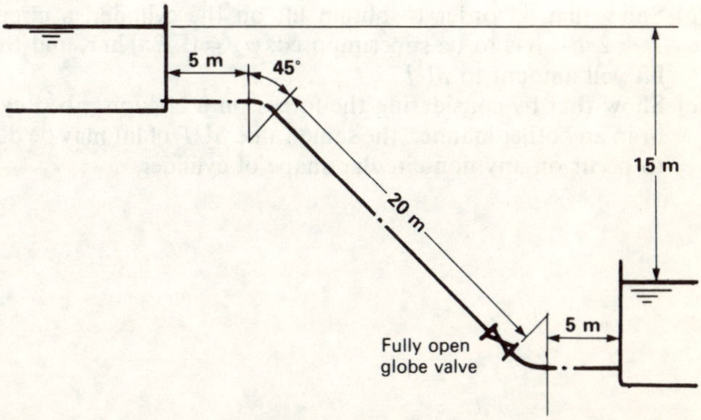

FIG. 86. *See* Question 4

5. Write an essay on "Frictional losses due to laminar and turbulent flow". Mention of relevant aspects of the work of Prandtl, Nikuradse, Moody and Colebrook–White should be included, together with descriptions of the various friction factor charts and their relative merits.

6. A pipeline which is to be laid between a pumping station and the intake to a supply network is to carry water at a rate of 0·6 cubic metres per second. Using the Moody diagram, *see* Fig. 84, estimate the required internal diameter d of the pipe if the overall length is 3 km, the relative roughness k/d is 0·03, the pressure head delivered by the pump is 70 m and the back pressure at the intake to the network is 200 kN/m². If a second pipe of the same diameter and material is connected into the first, part way along its length, runs parallel with it, and is required to increase the flow to the network by 30 per cent, at what distance along the first pipe from the intake must the tapping be made?

7. Describe with the aid of sketches the distribution of soot particles emitted at the top of a chimney 100 m high, at the rate of 1000 particles per m³ of flue gas rising up the 4-metre diameter chimney at a velocity of 4 m/s. The average soot particle has a terminal velocity in air of 5 mm/s.

Assume the ground to be effectively a rough plane over which the air flows in turbulent motion with an average velocity at the top of the chimney of 5 m/s, with winds predominantly from the west.

(*Note*: there is no exact answer to a question of this nature. Some simplifying assumptions have to be made about the extent of lateral spread of the soot particles, where they settle, and to what height the particles in the centre of the chimney are carried by the hot flue gases before they start falling.)

8. (*a*) Determine the equation for the streamlines of potential flow around a circular cylinder of radius a by combining a doublet with a uniform flow of velocity U. The doublet consists of a source, $\psi_1 = \theta Q/(2\pi)$, and sink, $\psi_2 = \theta Q/(2\pi)$, in which the distance d between source and sink vanishes as the value of Q increases in accordance with the relationship $Qd = \text{constant} = 2\pi U a^2$. The stream function of U is $\psi_3 = Uy$.

(*b*) Show that, in order to obtain lift on the cylinder, a circulation $\Gamma = 2\pi a V$ has to be superimposed, $\psi_4 = (\Gamma/2\pi) \ln r$, and that the lift will amount to $\rho U \Gamma$.

(*c*) Show that by considering the forces on a circumscribed cylinder, or in any other manner, the same value $\rho U \Gamma$ of lift may be deduced to occur on any non-circular shape of cylinder.

CHAPTER 12
FLUID MACHINES

ENERGY may exist in a number of forms. It may be associated with fluids, or with machinery in motion, or with electrical supply, etc. One of the major engineering achievements which revolutionised living standards in this century was the development of machines which *transform* energy from one form to another (*e.g.* mechanical to electrical, heat to mechanical, etc.).

Most fluid machines are designed for energy transformation and they may be divided into three broad categories:

1. Machines which transform an input of electrical or mechanical energy into an output of fluid energy (*e.g.* pumps, fans, compressors).
2. Machines which transform an input of fluid energy into an output of mechanical or electrical energy (*e.g.* water turbines).
3. Devices which simply transfer energy, say from a motor to drive shaft (*e.g.* hydraulic couplings).

A distinction may also be drawn between "positive displacement" and "rotodynamic" machines. In a positive displacement pump or compressor fluid is drawn into an enclosed space (for example a cylinder) which is then mechanically sealed. The fluid is pressurised by a piston and expelled, and the process is then repeated. Flow is thus intermittent in such machines. By contrast, in rotodynamic machines the fluid passes through continuously, energy transformation being effected by a rotating element. This chapter concentrates on rotodynamic machines.

ROTODYNAMIC MACHINES

Rotodynamic machines may be designed for duty as pumps, fans, turbines, or compressors. As stated above, these machines provide a continuous fluid flow. As the fluid passes through the rotating element a transfer of energy takes place and the fluid is then discharged from the machine.

In order to be able to estimate the rate of energy transfer to or from the fluid it is necessary to make use of velocity vector diagrams which represent the flow as the fluid enters and as it leaves the element. To show how this is done some typical rotodynamic machines will be analysed.

THE CENTRIFUGAL PUMP

A centrifugal pump has two main components, namely a rotating element known as an impeller, and a casing which completely encloses the impeller (see Fig. 87). The impeller may be cast or fabricated and essentially comprises three parts, namely a back plate or back shroud, a front shroud which has a central hole (or eye), and a series of vanes between. The impeller is attached to a drive shaft which passes out through a seal in the casing to the motor, which is usually electric. The suction and delivery pipelines are connected to their respective flanges on the casing. When the pump is in operation the liquid enters the casing and passes into the impeller through the eye. The impeller is rotating and this motion imparts energy to the liquid as it flows outwards through the impeller. The liquid is discharged from the outer periphery of the impeller into the outer part of the casing (the "volute"), which is a passage whose cross section gradually increases as it approaches the delivery passage. The function of the volute is to provide a smooth transition for the flow between the impeller periphery and the discharge pipeline. Losses do occur in the volute and some pumps incorporate a ring of fixed vanes (known as a diffuser ring) around the impeller to guide and decelerate the flow so that these losses may be minimised.

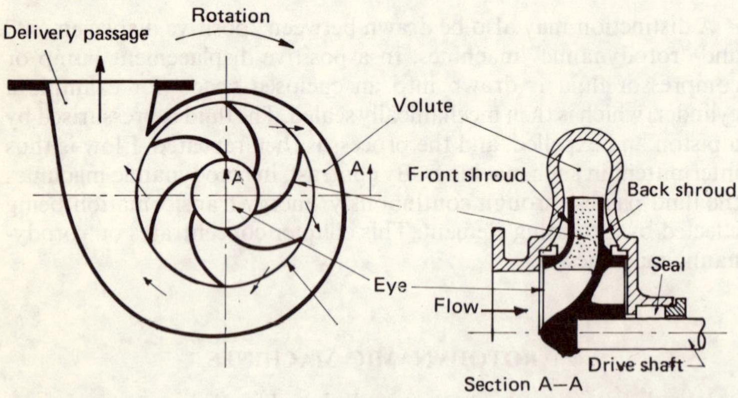

FIG. 87. The centrifugal pump—sectional views

The actual flow patterns which occur in an impeller are very complex. To produce vector diagrams it is, therefore, necessary to make simplifying assumptions, which nevertheless still give a reasonable approximation to the actual flow. The principal assumptions are:

1. That the flow through the impeller is in the plane of rotation only (*i.e.* velocities have only radial or tangential components).

2. That the velocity at a given radius is uniform;

and hence

3. That the flow is axisymmetric about the drive axis;
4. That the path of the relative velocity vector is congruent to the curve of the vane. (This assumption is the simplest both to visualise and to use at the initial stage, though it should be noted that some impellers are designed to operate with a small angle—the "angle of incidence"—between the vane and the relative path of the fluid.)

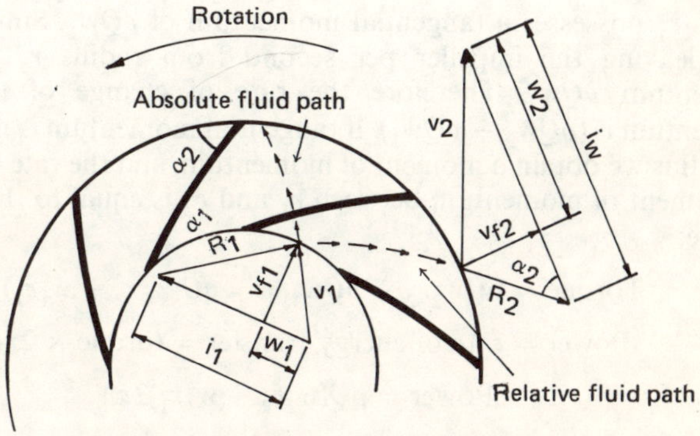

FIG. 88. Velocity diagrams for centrifugal pump

Figure 88 illustrates how the vector diagrams are constructed. The liquid enters the impeller in an axial direction and immediately turns so as to flow outwards in the plane of rotation. The liquid approaches the vane with absolute velocity v_1 (which is the velocity viewed from a stationary point).

If the impeller rotates at n rev/s, then its tangential velocity at the inner radius r_1 is $i_1 = 2\pi r_1 n$. The relative velocity R_1 between the impeller and the liquid is found by completing the vector triangle as shown. For the best flow pattern the vane angle α_1 at the inner radius should align with R_1. The liquid flows outwards in the passages between the vanes with its relative fluid path following the curvature of the vanes. It is discharged from the outer periphery of the impeller with relative velocity R_2, and at angle α_2. The vector representing the absolute tangential velocity of the impeller at the outer periphery (*i.e.* at r_2) is $i_2 = 2\pi r_2 n$. If i_2 is vectorially subtracted from R_2, the absolute fluid velocity v_2 is obtained. Two further velocity components are often useful. First there is the radial component of the absolute fluid velocity, v_{f1} (or v_{f2}). The other is the tangential component of the absolute fluid velocity w_1 (or w_2), often known as the "whirl" component.

The Euler equation for energy transfer

An expression for the energy ideally gained by the liquid from the impeller is now developed. The basis for the equation is an extension of the concepts of momentum already encountered in Chapter 4; however, only the acceleration in the tangential direction is considered. Now tangential momentum is the product of mass and tangential velocity, and since the tangential velocity of the fluid changes with radius the tangential momentum must change correspondingly. Thus the fluid entering the impeller per second at radius r_1 possesses a tangential momentum of $\rho Q w_1$. Similarly the fluid leaving the impeller per second from radius r_2 possesses momentum $\rho Q w_2$. Therefore the rate of change of tangential momentum is $(\rho Q w_2 - \rho Q w_1)$. If tangential momentum is multiplied by radius we obtain a moment of momentum, and the rate of change of moment of momentum between r_1 and r_2 is equal to the applied torque:

$$\text{Torque} = \rho Q w_2 r_2 - \rho Q w_1 r_1 = \rho Q(w_2 r_2 - w_1 r_1) \qquad (1)$$

$$\text{Power} = \text{rate of energy transfer} = \text{torque} \times 2\pi n$$

$$\therefore \ \text{Power} = \rho Q(w_2 r_2 - w_1 r_1)2\pi n \qquad (2)$$

but

$$2\pi r_1 n = i_1 \quad \text{and} \quad 2\pi r_2 n = i_2$$

$$\therefore \ \text{Power} = \rho Q(w_2 i_2 - w_1 i_1) \qquad (3)$$

Power per unit weight of fluid flowing is obtained by dividing the above expression by $\rho g Q$

$$\text{Power/unit weight} = H_i = \frac{\rho Q}{\rho g Q}(w_2 i_2 - w_1 i_1)$$

$$= \frac{1}{g}(w_2 i_2 - w_1 i_1) \qquad (4)$$

The dimensions of this expression are metres. H_i represents the ideal increase in the total head of the fluid due to the action of the impeller, *i.e.* the increase in head when all losses are ignored. This equation is known as Euler's equation.

CENTRIFUGAL FANS

The above remarks have centered on the centrifugal pump; however, the centrifugal fan operates in precisely the same way except that the fluid is a gas instead of a liquid. Providing that the density of the gas does not change significantly as it passes through the fan all the above equations are equally applicable to a fan as to a pump. The assumption of constant density is a reasonable one for a wide range of practical applications such as air conditioning systems, forced draught cooling towers, etc.

EXAMPLE 54

A centrifugal pump has an impeller whose dimensions are as follows: Diameter of "eye", 150 mm; impeller width at "eye", 80 mm. Outer diameter, 300 mm; width at O.D., 40 mm; vane outlet angle, $\alpha_2 = 25°$. For a flow of 150 litre/s of water and a speed of 1450 rev/minute estimate the following:

(*a*) vane entry angle assuming that v_1 is in a radial direction and therefore that $w_1 = 0$;
(*b*) theoretical impeller torque required;
(*c*) increase in head due to impeller;
(*d*) power supplied to fluid.

SOLUTION

(*a*) As v_1 is in a radial direction, clearly $v_1 = v_{f1}$. The discharge Q through the impeller is a function of the radial velocity component and of the cross-sectional area of the impeller in a plane perpendicular to radial velocity.

Thus at radius r_1, if vane thickness is neglected, the cross-sectional area perpendicular to flow will be $\pi \times$ diameter of eye $\times$ impeller width $= \pi d_1 b_1$.

$$\therefore\ Q = (\pi d_1 b_1) v_{f1} \quad \therefore\ v_1 = v_{f1} = \frac{Q}{\pi d_1 b_1} = \frac{0{\cdot}15}{\pi \times 0{\cdot}15 \times 0{\cdot}08} = 4 \text{ m/s}$$

$$i_1 = \pi d_1 n = \pi \times 0{\cdot}15 \times \tfrac{1450}{60} = 11{\cdot}39 \text{ m/s}$$

$$\therefore\ \alpha_1 = \tan^{-1}\left(\frac{v_{f1}}{i_1}\right) = \tan^{-1}\frac{4}{11{\cdot}39} \quad \therefore\ \underline{\underline{\alpha_1 = 19{\cdot}35°}}$$

(*b*) Theoretical torque is calculated using equation (1), but in order to evaluate it the value of w_2 must first be obtained. The cross-sectional area perpendicular to flow at $r_2 = \pi d_2 b_2$. Hence

$$v_{f2} = \frac{Q}{\pi d_2 b_2} = \frac{0{\cdot}15}{\pi \times 0{\cdot}3 \times 0{\cdot}04} = 4 \text{ m/s}$$

(Note that $v_{f1} = v_{f2}$; impellers are often designed to give constant v_f.)
From the vector diagram at the outer periphery,

$$w_2 = i_2 - \frac{v_{f2}}{\tan \alpha_2} = (\pi \times 0{\cdot}3 \times \tfrac{1450}{60}) - \frac{4{\cdot}0}{\tan 25°} = 14{\cdot}2 \text{ m/s}$$

Since $w_1 = 0$, torque $= \rho Q w_2 r_2 = 1000 \times 0{\cdot}15 \times 14{\cdot}2 \times 0{\cdot}15 = 319{\cdot}5$ N-m.

(*c*) Increase in head $= H_i = (1/g)(w_2 i_2 - w_1 i_1)$ (equation (4)).

$$\text{Since } w_1 = 0, \quad H_i = \frac{1}{g} w_2 i_2 = \frac{1}{9{\cdot}81} \times 14{\cdot}2 \times (\pi \times 0{\cdot}3 \times \tfrac{1450}{60})$$

$$\underline{\underline{= 33 \text{ m}}}$$

(*d*) Power supplied to fluid is calculated using equation (3), as $w_1 = 0$,

$$\text{Power} = \rho Q w_2 i_2 = 1000 \times 0{\cdot}15 \times 14{\cdot}2 \times (\pi \times 0{\cdot}3 \times \tfrac{1450}{60})$$

$$= \underline{\underline{48{\cdot}51 \text{ kW}}}$$

Effect of impeller vane angle

The above analysis of impeller performance indicates that the total energy (or head) transferred to the liquid is dependent on the whirl velocities w_1 and w_2. The velocities (including whirl) at the impeller inlet eye are generally determined by the layout of the system upstream of the pump or fan. However, at the impeller outlet the vane angle α_2 may be selected by the designer, and the magnitude of w_2 depends largely on α_2. For a given discharge and impeller size it is, therefore, theoretically possible to increase the transfer of energy to the fluid by increasing α_2. This increase will be in the form of kinetic energy since an increase in w_2 implies an increase in v_2 and, therefore, in $v_2^2/2g$. If v_2 is greater than the velocity in the discharge pipeline, then the flow has to decelerate in the volute and deceleration or "diffusion" processes usually involve shock losses. There is, therefore, no point in designing an impeller with an unduly large vane outlet angle, since much of the additional energy gained at the impeller is lost in the volute. For most centrifugal pumps α_2 falls within the range 15–90°, whereas for fans the range extends to 140°.

When the value of α_2 is less than 90°, the impeller is said to have a "backward" vane angle and if α_2 is greater than 90° it has a "forward" vane angle.

ENERGY LOSSES IN CENTRIFUGAL PUMPS AND FANS

The overall energy transfer actually achieved by pumps or fans is always less than that predicted by the Euler equation. The principal reasons for the losses are as follows:

1. It has already been shown that the fluid is accelerated tangentially as it passes outwards through the impeller. The impeller vanes must, therefore, apply a net tangential force to the fluid in the direction of acceleration. This implies that the pressure on the front of the vane must be higher than the pressure on the back (Fig. 89). This pressure difference also affects the relative velocity pattern, the relative fluid velocities close to the back of the vane being higher than the relative velocities close to the front of the vane. In other words, relative fluid velocity is *not* uniform at a given radius as assumed for the Euler equation, instead velocity varies across the passage between vanes as shown in Fig. 89. The relative velocity vector R_2, therefore, actually leaves at angle β_2 and not at the vane angle α_2. The actual

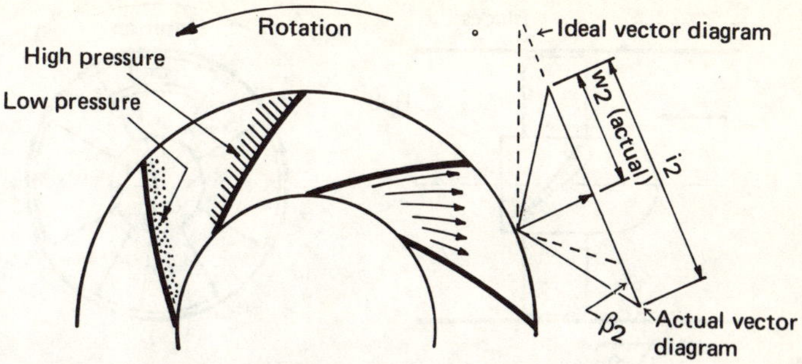

FIG. 89. Effect of pressure distribution on velocity diagram in centrifugal machine

whirl velocity is thus less than the ideal value and this results in a corresponding reduction in energy transfer to the fluid.

2. As the fluid passes through the impeller it is subject to both frictional and "shock" losses.

3. When the fluid leaves the impeller and flows around the volute there are further frictional and shock losses.

4. The fluid in the volute is under pressure, whereas at the entry to the eye pressure is low (often below atmospheric); consequently unless there is an extremely good seal between the volute and the impeller eye some of the high pressure fluid will leak back to the eye and recirculate through the impeller, thus wasting energy.

AXIAL FLOW PUMPS AND FANS

The rotating impeller element in an axial flow machine is in the form of a propeller. This is often mounted in a pipe or duct (Fig. 90). The fluid passes through the impeller from the intake to the delivery side, and it is usually assumed that any given particle of fluid remains at the same radial distance from the centre line throughout. The impeller consists of a number of blades which are attached to a central hub. The blades are at an angle to the axis and some machines incorporate controls to permit this angle to be varied. The hub is fixed to the drive shaft which transmits power to the impeller from the motor. The shaft runs in a bearing which is rigidly supported.

A section taken through a blade at a given radius shows that it is of aerofoil (aeroplane wing) form and indeed the relative flow pattern around the blades is similar in many ways to flow around the wing of an aircraft.

The Euler equation for axial flow machines

Since flow conditions may vary with radius, the velocities will first

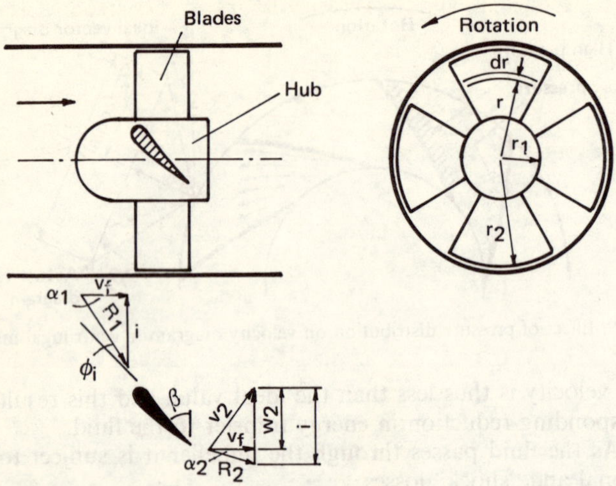

FIG. 90. Axial flow machine

be examined at one specific radius r. At this radius the tangential velocity of the impeller blade must be the same for the inlet or outlet side of the blade; thus, using the same notation as for the centrifugal impeller,

$$i_1 = i_2 = 2\pi rn = \text{constant} = i$$

The cross-sectional area of the flow passage is the same at the inlet or outlet. Also if the flow is incompressible, then the volumetric discharge Q remains constant. Since Q is the product of mean axial velocity of fluid and cross-sectional area, then the mean axial velocity must be the same for inlet or outlet; therefore it is reasonable to deduce that the local axial velocities (at radius r) must also be the same at inlet and outlet.

$$v_{f1} = v_{f2} = \text{constant} = v_f$$

If the fluid upstream of the impeller flows in an axial direction, then

$$w_1 = 0 \qquad \text{and} \qquad v_1 = v_f$$

The velocity diagram for the inlet is then simple to construct (Fig. 90).

The relative velocity vector approaches the blade at an angle ϕ_i known as the angle of incidence. For a given blade section shape there is an optimum value of ϕ_i at which losses are minimised. The blade section is at an angle β to the plane of rotation.

As the fluid passes over the blade it is accelerated tangentially, the magnitude of this acceleration being a function of blade section shape, and of the number of blades. At the outlet the relative velocity R_2 is at angle α_2 to the axis, and the velocity diagram is constructed as shown in the figure.

The magnitudes R_2 and α_2 must be such that $R_2 \cos \alpha_2 = v_f$.

Now if a discharge $\mathrm{d}Q$ passes axially through a ring-shaped element δr thick and of radius r, then the tangential force acting on that fluid must be equal to the rate of change of tangential momentum:

$$F = \rho\, \mathrm{d}Q w_2$$

$$\therefore \text{ Torque } T = F \times r = \rho\, \mathrm{d}Q w_2 r$$

$$\text{Power} = 2\pi n T = 2\pi n \rho\, \mathrm{d}Q w_2 r = \rho\, \mathrm{d}Q w_2 i$$

$$\therefore \text{ Power/unit weight of fluid} = \frac{\rho\, \mathrm{d}Q w_2 i}{\rho g\, \mathrm{d}Q} = \frac{1}{g} w_2 i$$

Now the above equations apply at one radius only since i certainly varies with radius and w_2 may also vary. In order, therefore, to estimate the performance of the impeller, it is necessary to integrate (or sum) between the hub (r_1) and tip (r_2). This is not as easy as it sounds since the real flow conditions are quite complex and w_2 may not be a simple function of r. Nevertheless it is possible to derive approximate solutions if simplifying assumptions are made. There are three alternatives which lead to mathematically straightforward solutions.

1. w_2 = constant;
2. w_2 varies with r to form a forced vortex;
3. w_2 varies with r to form a free vortex.

Which of the above gives the closest approximation for a particular impeller depends on the shape of the blade section, on the variation of β with radius, and on the proximity of one blade to another (*i.e.* on the number of blades and the blade thickness). A series of aerofoil section blades such as those on pumps or fans is known as a "cascade". A study of cascade theory is beyond the scope of this book. A considerable amount of data is available to the designer covering cascades with different blade sections and blade spacing.

EXAMPLE 55

An axial flow fan rotates at 725 rev/min and delivers 15 m³/s of air for a ventilation system. The hub diameter is 0·5 m and the tip diameter is 1·0 m. As the air passes from inlet to outlet the relative velocity vector is deflected from angle α_1 to α_2, and at 0·375 m radius the change of angle $(\alpha_1 - \alpha_2) = 30°$. Estimate:

1. Power ideally supplied to the fluid;
2. Ideal head rise across the impeller.

In order to evaluate the outlet whirl velocity, w_2, at any radius, assume forced vortex flow downstream of the impeller. Assume also that the air density remains constant, at 1·28 kg/m³.

SOLUTION

Axial velocity

$$v_f = \frac{Q}{A} = \frac{15{\cdot}0}{\frac{\pi}{4}(1{\cdot}0^2 - 0{\cdot}5^2)} = 25{\cdot}46 \text{ m/s}$$

(assuming negligible tip clearance, *i.e.* duct diameter = 1·0 m)

$$\text{tangential velocity of impeller at } 0{\cdot}375 \text{ m radius} = 2\pi \times 0{\cdot}375 \times \tfrac{725}{60} = 28{\cdot}47 \text{ m/s}$$

$$\therefore\ \alpha_1 = \tan^{-1}\left(\frac{28{\cdot}47}{25{\cdot}46}\right) = 48{\cdot}19°$$

Since $\alpha_1 - \alpha_2 = 30°$, $\alpha_2 = 48{\cdot}19 - 30 = 18{\cdot}19°$. v_f = constant, ∴ at impeller blade outlet, and at 0·375 m radius, $w_2 = i - v_f \tan\alpha_2 = 28{\cdot}47 - 25{\cdot}46 \times \tan 18{\cdot}19° = 20{\cdot}1$ m/s.

For a forced vortex angular velocity ω = constant at any radius, and hence

$$\omega = \frac{w_2}{r} = \frac{20{\cdot}1}{0{\cdot}375} = 53{\cdot}61 \text{ rad/s}$$

Now to estimate the performance of the impeller as a whole, velocities must be expressed as functions of r:

$$w_2 = \omega r = 53{\cdot}61r \text{ m/s}$$

$$i = 2\pi rn = 2\pi r(\tfrac{725}{60}) = 75{\cdot}92r$$

Discharge dQ through a ring-shaped element

$$= v_f \times (\text{area of ring})$$

$$= v_f \times 2\pi r\,\mathrm{d}r$$

$$= 25{\cdot}46 \times 2\pi r\,\mathrm{d}r = 159{\cdot}97r\,\mathrm{d}r$$

$$\therefore\ \text{Ideal power supplied} = \int_{0{\cdot}25}^{0{\cdot}5} \rho\,\mathrm{d}Q w_2 i$$

$$= \int_{0{\cdot}25}^{0{\cdot}5} 1{\cdot}28 \times 159{\cdot}97r\,\mathrm{d}r \times 53{\cdot}61r \times 75{\cdot}92r$$

$$= \int_{0{\cdot}25}^{0{\cdot}5} 833\,394{\cdot}29 \times r^3\,\mathrm{d}r = 12\,207{\cdot}9 \text{ W}$$

say <u>12·208 kW</u>

$$\text{Ideal head} = \frac{\text{power}}{\rho g Q} = \frac{12208}{1{\cdot}28 \times 9{\cdot}81 \times 15{\cdot}0} = \underline{64{\cdot}8 \text{ m head of air}}$$

Use of stators in axial pumps and fans

Some of the more sophisticated fans or pumps incorporate a set of stationary blades (known as a stator) just downstream of the impeller. The stator blade angle is designed to "straighten" the flow so that when the fluid leaves the stator it is flowing in the axial direction, *i.e.* so that its whirl velocity is zero and the final absolute velocity of fluid v_3 is equal to v_f.

Now as $v_2 > v_3$, it follows that there must be a corresponding reduction in kinetic energy, and applying Bernoulli's equation we know that a reduction in kinetic energy must be compensated by a corresponding increase in pressure. Thus the effect of a stator is to improve overall pressure rise attainable by the fan.

Losses occur both in the impeller and stator, so the ideal performance figures are never attained in practice.

PERFORMANCE DATA FOR PUMPS AND FANS

Although approximate estimates of the performance of a pump or fan may be obtained from theory, all the performance data published by manufacturers are derived from laboratory tests. The principal parameters of performance are: discharge, Q (or mass flow $\dot{m}$); the actual head H imparted to the fluid, and the power P supplied to the drive shaft of the machine. From these two further parameters may be derived: namely, the power supplied to fluid ($= \rho g Q H$) and the efficiency η ($= \rho g Q H / P$). A machine is tested over its range of discharge and the performance data may then be plotted. Typical performance curves (characteristic curves) are shown in Fig. 91. The power (P), head (H) and efficiency (η) are plotted to a base of discharge.

The following points should be noted:

1. For axial flow and backward-angled centrifugal machines there is a maximum power requirement. The drive motor should, therefore, be adequately rated for that maximum.
2. Radial (90°) and forward-angled centrifugal impellers generate a greater total head, but the power requirement increases with discharge. It will be shown in the next section that the actual operating discharge of a machine is a function of both the machine and the system in which it is installed. The characteristics of a complex pipe or duct system are not easily predicted with precision—nor, therefore, is the operating discharge. Selection of a motor for a machine with a radial or forward-angled impeller therefore requires very great care.
3. Axial flow machines are usually employed where a high

discharge is required. They do, however, suffer from the following disadvantages:

(*a*) Their suction capability is limited.
(*b*) The efficiency curve descends steeply on either side of the maximum, the economic operating range is therefore narrow.
(*c*) When discharge falls below the optimum value, the head discharge characteristic reveals a region of potential operational instability (this also applies to some centrifugal machines).
(*d*) The pressure rise attainable by an axial machine is less than for a corresponding centrifugal machine. This can be offset by the use of multi-stage systems (*i.e.* two or more machines in series).

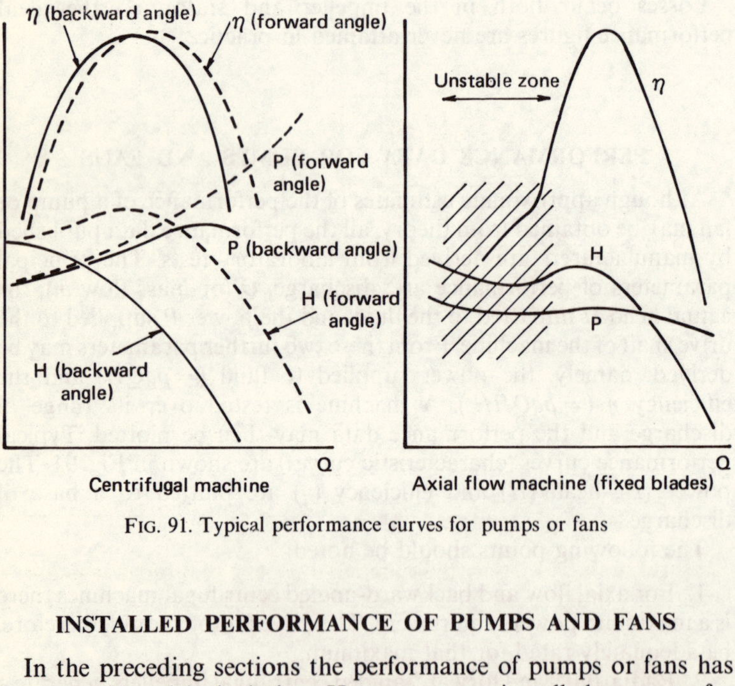

FIG. 91. Typical performance curves for pumps or fans

INSTALLED PERFORMANCE OF PUMPS AND FANS

In the preceding sections the performance of pumps or fans has been considered in isolation. However, once installed as part of a system, the performance of the pump or fan and the characteristics of the system are interdependent. The relevant characteristics of a typical pump and pipeline system are illustrated in Fig. 92.

For a fluid to flow it must be supplied with sufficient energy. Thus for flow in the pipe on the suction side of the pump (connecting the sump tank to the pump) energy is required:

1. to raise the liquid through height z_1;

2. to provide kinetic energy $u_1^2/2g$;
3. to overcome energy loss h_{f1} due to friction and "shock".

The sum of these three components is H_{m1}, the manometric suction head. On the delivery side the pipeline diameter is often smaller than the suction pipe and then $u_2 > u_1$.

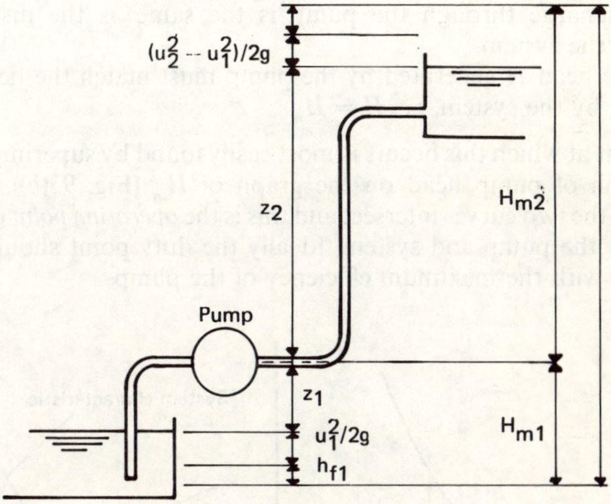

FIG. 92. Principal characteristics of a pipeline system

Therefore for flow in the discharge pipe, energy is required.

1. to raise the liquid through height z_2;
2. to provide additional kinetic energy $(u_2^2 - u_1^2)/2g$;
3. to overcome energy loss h_{f2}.

The sum of these three components is H_{m2}, the manometric delivery head. The total energy which must be provided in order to induce a flow is therefore

$$
\begin{aligned}
H_m &= H_{m1} + H_{m2} \\
&= (z_1 + z_2) + (h_{f1} + h_{f2} + u_1^2/2g + (u_2^2 - u_1^2)/2g) \\
&= (z_1 + z_2) + (h_{f1} + h_{f2} + u_2^2/2g)
\end{aligned}
$$

For a particular system z_1 and z_2 are constant, but the remaining quantities are all dependent on the magnitude of the square of the velocity of flow. It is usually more convenient to express these remaining quantities in terms of Q^2 rather than velocity squared. The above equation may then be written:

$$H_m = (z_1 + z_2) + KQ^2 \qquad (5)$$

where $$KQ^2 = (h_{f1} + h_{f2} + u_2{}^2/2g)$$

Equation (5) is the characteristic equation for the system; this may be presented in graphical form (Fig. 93(*a*)).

Now if the energy required is being provided by a pump it is clear that:

1. Discharge through the pump is the same as the discharge through the system.
2. The head H generated by the pump must match the head H_m required by the system, *i.e.* $H = H_m$.

The point at which this occurs is most easily found by superimposing the graph of pump head on the graph of H_m (Fig. 93(*b*)). When $H = H_m$ the two curves intersect and this is the *operating point* or *duty point* for the pump and system. Ideally the duty point should also coincide with the maximum efficiency of the pump.

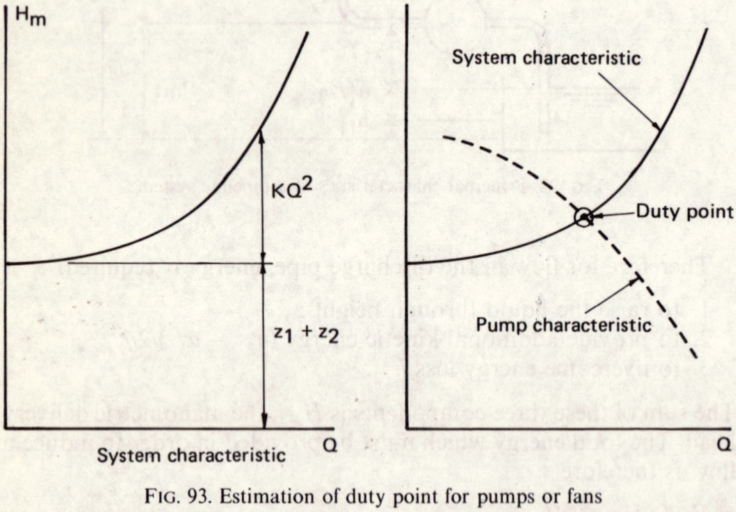

FIG. 93. Estimation of duty point for pumps or fans

The hydraulic characteristics of some systems may necessitate the use of more than one machine. Two configurations are possible:

(*a*) Series operation, in which the delivery flange of one machine is coupled to the intake flange of the next machine. The same discharge therefore passes through every machine. However, the total head imparted to the fluid is equal to the sum of the heads for each individual machine at the given discharge (*i.e.* $H = H_1 + H_2 + \cdots$).

(*b*) Parallel operation, in which a single main is split into a number of parallel lines (one for each machine) at the point where the machines are mounted. Every machine therefore operates at the same head. However, the total discharge is equal to the sum of the discharges for each machine at the given head (*i.e.* $Q = Q_1 + Q_2 + \cdots$).

The combined characteristic for a multi-machine arrangement may be constructed by "adding" the individual machine characteristics as outlined above. The combined characteristic should cross the system characteristic at one point only, otherwise unstable conditions may arise.

Most of the remarks in this section apply in principle equally to fans or pumps. For fans, however, the elevation (z_1 or z_2) is generally insignificant and the system characteristic is related solely to the kinetic energy and loss components, *i.e.* to the "KQ^2" term.

WATER TURBINES

Modern water turbine designs emerged in the late nineteenth century, and since then they have been installed in many parts of the world. Compared with electrical generation based on coal, oil or nuclear fuel, hydro-electric power has two major advantages. Firstly it does not produce pollutants, and secondly the energy source (water) is renewable as long as there is adequate rainfall.

The rotating element in a water turbine is known as a "runner". The water turbine requires an input of water which has a high energy content. The usual source of water is a reservoir, though some modern schemes have been designed to utilise tidal flows. As the water flows through the turbine, the runner transforms the fluid energy into a torque which is used to drive a generator.

There are three basic types of water turbine: the Francis, the axial flow and the Pelton; these are now analysed through worked examples.

The Francis turbine

This turbine design is used where there is a plentiful water supply but only a moderate static pressure head. The water is fed through a pipeline to the turbine casing, which appears similar to the casing for a centrifugal pump (*see* Fig. 94). For the turbine, however, the water is usually fed *into* the outer (volute) section of the casing, whence it flows *inwards* through the runner. Thus the flow and the energy transfer process in the turbine is the *reverse* of that for a pump. Another difference is that the turbine casing always incorporates a ring of guide vanes between the volute and the runner. The angle of the guide vanes is adjustable so as to provide the optimum direction of flow for the runner under varying conditions of flow or of power output.

As the flow passes through the runner the curvature of the runner vanes causes a change in its direction with a corresponding change in

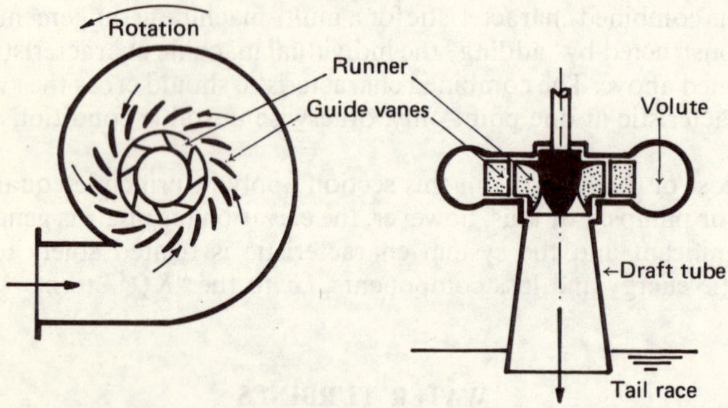

FIG. 94. Francis turbine

the tangential momentum. The reaction to this change of angular momentum is a torque which acts on the runner, providing the motive force for the turbine. The water is discharged through the eye into a "draft tube" and is usually drained away into an open channel or "tail race". The draft tube is tapered as shown so as to minimise the loss of kinetic energy in the discharge to the tail race. The system is designed to run full of water.

EXAMPLE 56

A Francis type reaction turbine has a runner having an outer diameter of 1200 mm and an inner "eye" diameter of 600 mm. The circumferential cross section is the same for the inlet and outlet of the runner. The guide vanes are set to an angle of 15° and the runner outlet angle is 25°. Given that the water enters the turbine with a total head of 25·5 m, the discharge is 6·6 m^3/s, the runner speed is 240 rev/min and the power output is 1·15 MW, estimate:

(*a*) the optimum angle of the runner vanes at entry;
(*b*) the kinetic energy head at the entry to the draft tube;
(*c*) the mechanical and overall efficiency of the turbine;
(*d*) the hydraulic efficiency.

Assume that there is zero whirl velocity at runner outlet. Figure 95 illustrates the construction of the velocity diagrams.

SOLUTION

(*a*) In order to calculate the optimum blade angle at entry we need to know the angle of the relative velocity vector R_1, and for this we must first find v_{f1}. Since we have insufficient information to complete the vector diagram for the

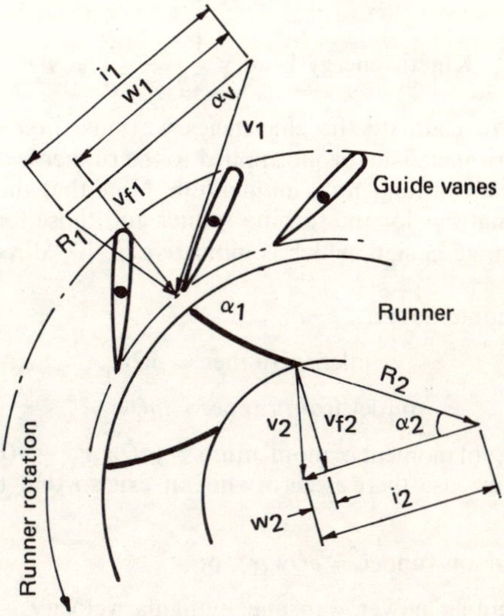

FIG. 95. Velocity diagrams for Francis turbine

inlet we will consider the vector diagram at outlet first. Here we may calculate the tangential velocity:

Outlet tangential velocity $= i_2 = \pi d_2 n = \pi \times 0{\cdot}6 \times \frac{240}{60} = 7{\cdot}54$ m/s

Vane angle at outlet = 25°

$\therefore$ Radial velocity $v_{f2} = 7{\cdot}54 \times \tan 25° = 3{\cdot}516$ m/s

As there is zero whirl $v_{f2} = v_2$.

Since circumferential area at outlet and at inlet are equal (*i.e.* cross-sectional area for flow is constant), we know that $v_{f2} = v_{f1}$; we may now solve the inlet vector diagram.

$$\text{Whirl velocity at inlet} = w_1 = \frac{v_{f1}}{\tan \alpha_v} = \frac{3{\cdot}516}{\tan 15°} = 13{\cdot}12 \text{ m/s}$$

Inlet tangential velocity $= \pi d_1 n = \pi \times 1{\cdot}2 \times \frac{240}{60} = 15{\cdot}08$ m/s

$$\therefore\ i_1 - w_1 = 15{\cdot}08 - 13{\cdot}12 = 1{\cdot}96 \text{ m/s}$$

$$\therefore\ \tan^{-1}\frac{3{\cdot}516}{1{\cdot}96} = 60{\cdot}9°,\ \underline{\underline{\text{say } 61°}}$$

$\therefore$ if runner vane entry angle $\alpha_1 = 61°$ the relative velocity vector and blade entry angle are in alignment.

(*b*) The diameter at the entry to the draft tube will be 600 mm and the velocity of flow is therefore

$$V = \frac{\text{discharge}}{\text{area}} = \frac{6{\cdot}6}{(\pi/4) \times 0{\cdot}6^2} = 14 \text{ m/s}$$

$$\therefore \text{ Kinetic energy head} = \frac{V^2}{2g} = \frac{14^2}{2g} = \underline{\underline{9{\cdot}99 \text{ m}}}$$

(*c*) In order to estimate the efficiencies we must first calculate power supplied to the runner. The torque applied to the runner is due to the rate of change of moment of tangential momentum. Note that the only difference between the equations for the turbine runner and those for the centrifugal pump is the change of sign, which is indicative of the "direction" of energy transfer.

Tangential momentum at:

$$\text{inlet to runner} = \rho Q w_1$$

$$\text{outlet from runner} = \rho Q w_2$$

$\therefore$ Rate of change of moment of momentum $= (\rho Q w_1 r_1 - \rho Q w_2 r_2) =$ torque. In this particular case there is zero whirl at exit so that the second term equates to zero.

$$\text{Torque on runner} = \rho Q w_1 r_1 \text{, or}$$

$$\begin{aligned} \text{Runner power} &= \text{torque} \times \text{angular velocity} \\ &= \rho Q w_1 r_1 \times 2\pi n = \rho Q w_1 i_1 \\ &= 1000 \times 6{\cdot}6 \times 13{\cdot}12 \times 15{\cdot}08 = 1{\cdot}306 \text{ MW} \end{aligned}$$

$$\begin{aligned} \text{Mechanical efficiency} &= \frac{\text{power output}}{\text{runner power}} = \frac{1{\cdot}15}{1{\cdot}306} \\ &= 0{\cdot}88 = 88\% \end{aligned}$$

$$\text{Overall efficiency} = \frac{\text{power output}}{\text{energy supplied}}$$

The energy or power to the turbine is provided by the water supply, and as the water enters the turbine with a total head of 25·5 m the energy supplied is

$$E = \rho g Q h = 1000 \times 9{\cdot}81 \times 6{\cdot}6 \times 25{\cdot}5 = 1{\cdot}651 \text{ MW}$$

$$\therefore \text{ Overall efficiency} = \frac{1{\cdot}15}{1{\cdot}651} = 0{\cdot}697 = \underline{\underline{69{\cdot}7\%}}$$

(*d*) Lastly

$$\text{Hydraulic efficiency} = \frac{\text{runner power}}{\text{energy supplied}} = \frac{1{\cdot}306}{1{\cdot}651} = 0{\cdot}791 = \underline{\underline{79{\cdot}1\%}}$$

The axial flow turbine

The axial flow turbine is appropriate where there is a substantial supply of water but the static pressure head is low. The turbine runner is similar in form to the impeller of the axial flow pump, except that the blade angle is designed so that the runner is driven by the water. The blade angle may be fixed but the efficiency of such machines falls off sharply on either side of the optimum so that the operating range is narrow.

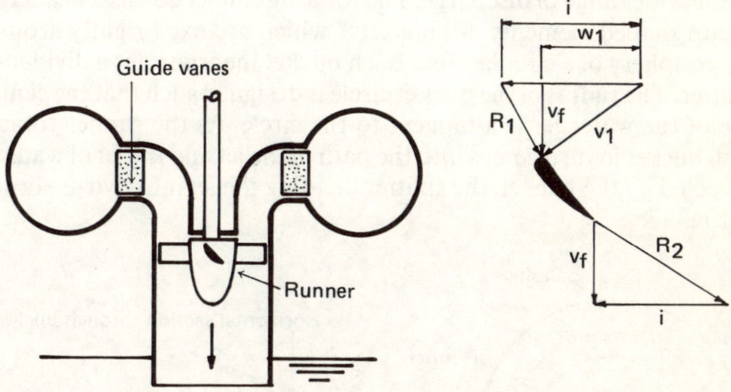

FIG. 96. Axial flow turbine

For this reason many axial flow turbines incorporate a mechanism for varying the blade angle, so that efficiency may be maintained over a wider range of discharge and power output. These are known as "Kaplan" turbines. Upstream of the runner is a ring of guide vanes. The angle of the vanes is adjustable, and the vanes fulfil the same function in axial flow turbines as in Francis turbines. Some recent designs of the Kaplan type are able to operate as a pump or as a turbine and these are used in "pumped storage" schemes. Such schemes utilise electricity at night when the demand is low, to pump water up to a storage reservoir. During the day when electrical demand is high the stored water is released, the runner is adjusted to run as a turbine and electricity is generated.

An approximate estimate of the performance of an axial flow turbine may be attempted on the same theoretical basis as was used for the example on the axial flow fan. Due to the guide vanes the *incoming* water has a whirl component w_1, whilst ideally the flow leaves the runner with zero whirl, which in effect is the reverse of the velocity pattern for the pump or fan. A more precise prediction of performance would involve the use of cascade theory.

The Pelton wheel

This type of machine is used where the available water supply has a high static pressure head. The water is fed from the reservoir through the pipeline system to the turbine casing. The pipeline terminates in one or more nozzles which are inside the casing and which convert the pressure energy into the kinetic energy of a high velocity jet of water. The flow rate is controlled by a streamlined valve which is incorporated in each nozzle and which maintains a high jet velocity over a wide range of discharge. The rotating runner consists of a series of cup shaped elements or "buckets" which are fixed rigidly around the periphery of a circular disc. Each bucket incorporates a divider or splitter. The radius of the bucket circle is designed such that the centre line of the water jet is a tangent to the circle. As the runner rotates each bucket in turn comes into the path of the jet and the jet of water is deflected by the bucket, the splitter dividing the jet into two as shown in Fig. 97.

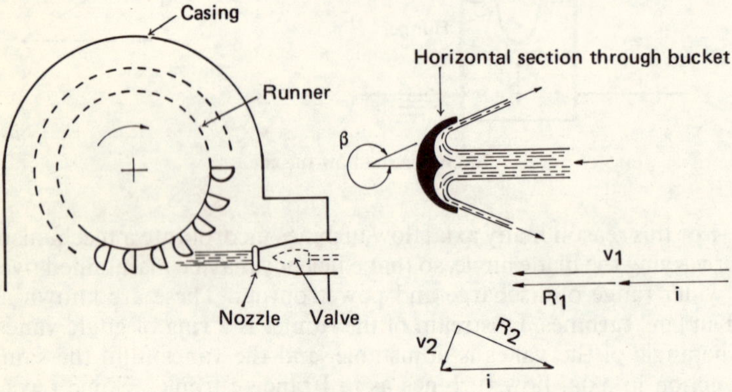

FIG. 97. The Pelton wheel

After striking the bucket the water falls freely downwards and is drained away. The casing of a Pelton wheel is only fitted to prevent undue splashing, and is not intended to run full of water; the wheel thus operates at atmospheric pressure.

EXAMPLE 57

The water supply for a hydro-electric power station comes from a reservoir whose water level is 300 m above the turbine. The pipeline is 750 mm diameter; its friction factor is 0·007 and it is 5 km long. The turbine is of the Pelton type and the nozzle is 100 mm diameter. The buckets are arranged at an effective diameter of 1.70 m and deflect the jet through 165°. As the water flows over the bucket friction reduces the velocity of the water by 12%. Estimate the hydraulic and overall efficiency of the turbine if mechanical losses account for 10% of the turbine power output, and the wheel rotates at 380 rev/min. At what wheel speed would maximum efficiency be attained?

SOLUTION

The available head of water must overcome friction in the pipeline and provide the kinetic energy of the jet. If the velocity of the jet is v_1, then:

$$300 \text{ m} = \frac{0{\cdot}33 fLQ^2}{d^5} + \frac{v_1{}^2}{2g}$$

But
$$v_1 = \frac{Q}{\text{nozzle area}}$$

$$\therefore\ 300 \text{ m} = \frac{0{\cdot}33 \times 0{\cdot}007 \times 5000Q^2}{0{\cdot}75^5} + \frac{1}{2g}\frac{Q^2}{((\pi/4) \times 0{\cdot}1^2)^2}$$

$$\therefore\ Q = 0{\cdot}5855 \text{ m}^3/\text{s}$$

$$v_1 = 74{\cdot}55 \text{ m/s}$$

Tangential velocity of bucket,

$$i = \pi \times 1{\cdot}7 \times \tfrac{380}{60} = 33{\cdot}82 \text{ m/s}$$

From the vector diagram above, as jet enters bucket

$$R_1 = 74{\cdot}55 - 33{\cdot}82 = 40{\cdot}73 \text{ m/s}$$

which is the initial velocity relative to the bucket. The water loses 12% of its velocity due to friction; therefore, as jet leaves bucket, $R_2 = 40{\cdot}73 \times 0{\cdot}88 = 35{\cdot}84$ m/s. The jet has also been deflected through 165°.

The thrust on the bucket is due only to the change of momentum of the water in the tangential direction (*i.e.* in line with the plane of rotation of the impeller). Components of momentum perpendicular to the plane of rotation have no effect on power output.

Final tangential component of velocity of water $= R_2 \cos\beta = 35{\cdot}84 \times \cos 165° = -34{\cdot}62$ m/s (the negative sign indicates that the direction of this component is opposite to the direction of the initial relative velocity).

F = Effective force on bucket = rate of change of tangential momentum,

$$\therefore\ F = \rho Q(R_1 - R_2 \cos\beta)$$
$$= 1000 \times 0{\cdot}5855(40{\cdot}73 - (-34{\cdot}62)) = 44\,117 \text{ N}$$

Runner power = Effective force on bucket × bucket velocity

$$= 44117 \times 33{\cdot}82 = 1{\cdot}492 \text{ MW}$$

Mechanical losses account for 10% of power

$\therefore$ Power output from turbine $= 0{\cdot}9 \times 1{\cdot}492 = 1{\cdot}343$ MW

Kinetic energy supplied by water jet $= \tfrac{1}{2}\rho Q \times v_1{}^2$

$$= \tfrac{1}{2} \times 1000 \times 0{\cdot}5855 \times 74{\cdot}55^2 = 1{\cdot}627 \text{ MW}$$

$$\therefore\ \text{overall efficiency} = \frac{1{\cdot}343}{1{\cdot}627} = \underline{\underline{82{\cdot}5\%}}$$

$$\text{and hydraulic efficiency} = \frac{1{\cdot}492}{1{\cdot}627} = \underline{\underline{91.7\%}}$$

Now $R_1 = (v_1 - i)$ and $R_2 = k(v_1 - i)$, where k = Loss coefficient ($k = 0{\cdot}88$ for this turbine). Thus the hydraulic efficiency may be expressed as:

$$\eta_H = \frac{\rho Q(R_1 - R_2 \cos \beta)i}{\frac{1}{2}\rho Q v_1^{\,2}} = \frac{(v_1 - i)(1 - k \cos \beta)i}{v_1^{\,2}/2}$$

It is easily shown (graphically or by differentiation) that η_H *is a maximum when* $v_1 = 2i$. $\therefore\ i = 74{\cdot}55/2 = 32{\cdot}275$ m/s and the corresponding wheel speed

$$= \frac{32{\cdot}275 \times 60}{\pi \times 1{\cdot}7} = \underline{\underline{362{\cdot}6 \text{ rev/min}}}$$

INSTALLED PERFORMANCE OF TURBINES

The principles of interaction between a fluid machine and a pipe system have been covered in detail for pumps. The only essential difference when considering turbines is that the direction of flow (and of energy transfer) is reversed. Thus the energy supply ideally available to drive the turbine is the potential energy z. However, there are losses in the system, due to pipe friction, etc., so these have to be *deducted* from z to give the manometric head H_m which is the head actually available at the turbine (Fig. 98).

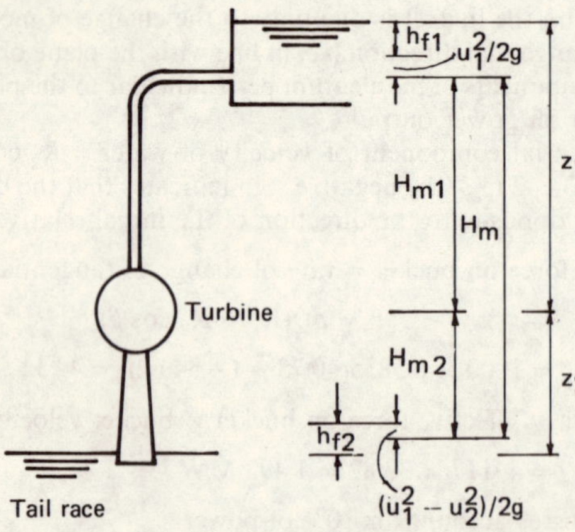

FIG. 98. Characteristics of supply system for turbine

For the Pelton wheel the effective head is the head available above the nozzle only, *i.e.* $H_{m1} = (z_1 - \text{losses})$. For axial flow and Francis machines, however, there is a continuous pipe system from the supply reservoir to the tail water so that the manometric heads are

$$H_{m1} = z_1 - h_{f1} - u_1^{\,2}/2g$$

$$H_{m2} = z_2 - h_{f2} - \frac{(u_2{}^2 - u_1{}^2)}{2g}$$

(N.B. usually $u_1 > u_2$ in turbines.)

CAVITATION

If the pressure of a liquid falls sufficiently far below atmospheric pressure, bubbles of vapour begin to form, and the liquid may thus be made to boil even at normal air temperatures. This phenomenon is known as cavitation and it is of great significance in the operation of fluid machines. The absolute pressure P_{vap} at which cavitation commences is a function of the temperature of the liquid and is also affected by the presence of dissolved gas in the liquid. Regions of low pressure may occur in a fluid machine. For example, when pumps have to raise a liquid through height z_1 from a sump tank (Fig. 92), the pressure at the pump inlet may fall below atmospheric pressure. The *negative* value of gauge pressure head at which cavitation commences is

$$H_c = \frac{(P_{atm} - P_{vap})}{\rho g}$$

The difference between the head at the pump intake and the vapour pressure is known as Net Positive Suction Head, and thus

$$\text{NPSH} = H_c - H_{ms}$$

A German engineer, D. Thoma, suggested a cavitation parameter σ_T which relates NPSH to total pump head, H_m:

$$\sigma_T = \frac{\text{NPSH}}{H}$$

(for turbines $\sigma_T = (H_c - z)/H_m$, where z = height of turbine above tail race).

Pumps and turbines may be tested for their cavitation characteristics and a critical value of σ_T is then calculated. Cavitation can cause noise and vibration. It can also damage the impeller or runner of the machine by progressive erosion.

ROTARY MACHINES

There are a number of important industrial applications for which the centrifugal or axial pumps are unsuitable. Typical of such cases are hydraulic power transmission systems, in which a pump and an hydraulic drive unit are connected by a closed circuit. The requirement here is for a pump which is compact and which is capable of giving a very high pressure (20 MN/m^2 or more) at a low or moderate discharge. Rotary pumps are very often used for such duties. Strictly speaking, rotary pumps belong to the "positive displacement" category, even though they achieve flows which appear to be almost continuous. Two typical examples are shown in Fig. 99.

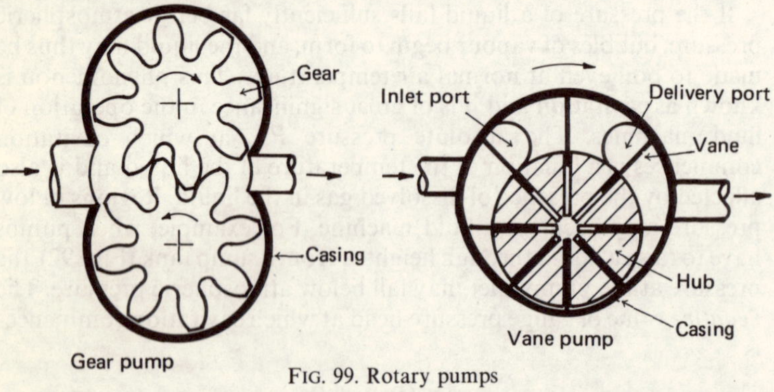

FIG. 99. Rotary pumps

The gear pump consists of two (or more) meshing gear wheels mounted on parallel shafts of which one shaft is motor driven. The wheels are enveloped in a casing. As the wheels rotate the incoming liquid is entrained in the space between the gear teeth and is then carried around to the delivery port. Thus the theoretical capacity of the pump is

$$Q = tvn$$

where t = total number of teeth on gear wheels
v = volume enclosed between one pair of teeth
n = rotational speed.

Gear pumps may be operated at pressures up to about 20 MN/m^2 and give discharges of up to 1·6 litres/revolution.

The vane pump incorporates a hub with a number of radial slots. A vane is mounted in each slot and the vane is free to slide radially. The hub is fixed on a shaft which is mounted eccentrically in the surrounding casing. When the shaft rotates the vanes are so disposed that their tips are in sliding contact with the inner wall of the casing in order to form a seal. As an adjacent pair of vanes pass the inlet port, the volume enclosed between them is increasing due to the effect of the eccentricity. Liquid is therefore drawn into the pump, and is then carried round towards the delivery port. As the vanes pass the delivery port the eccentricity causes the inter-vane volume to progressively decrease, forcing the liquid out into the delivery pipe.

The maximum working pressure is about 7 MN/m^2.

Theoretically the discharge of a rotary pump increases with rotational speed regardless of the increase of head across the pump. In practice it is impossible to obtain a perfect sealing arrangement between the high and low pressure sides of the pump, and some fluid therefore leaks back past the tips of the vanes or gear teeth. This problem of leakage becomes more severe at higher operating pressures. Another important factor is that the liquid must possess lubricating qualities, since there is sliding contact between gear teeth or between the vane tips and the casing. The liquid must also be filtered since any solid particles in suspension in the liquid can cause damage by scoring the surfaces in sliding contact.

Because these machines are positive displacement units, the liquid is actually discharged as a sequence of discrete volumes. In consequence the delivery pressure is not constant but pulsates slightly. With minor modifications either of the above machines may be used "in reverse" to run as a drive unit, *i.e.* as a "motor" or "turbine".

Hydraulic power transmission systems are often more compact, lighter and cheaper than electrical equivalents. They are also more flexible in operation than most mechanical systems, although efficiency is often lower. The pump may be driven by a diesel or petrol motor and hydraulic power circuits have, therefore, been used extensively for vehicle mounted systems, *e.g.* farm machinery and heavy earthmoving plant. Gear and vane pumps are also used in the petrochemical and process industries and in certain aspects of food processing.

QUESTIONS

1. A centrifugal fan is designed to deliver 3 m^3/s of air (air density = 1·28 kg/m^3), when rotating at 1450 rev/min. The impeller characteristics are as follows:

Inlet diameter = 400 mm
Outlet diameter = 700 mm
Width of flow passage at inlet = 200 mm

The flow passage is designed to give constant radial velocity (v_f). The blades are backward angled, the exit angle being 45°.

If the flow enters the fan with zero whirl, determine:

(*a*) the inlet angle of the vane;
(*b*) the theoretical torque;
(*c*) the theoretical head in metres of air and mm of water.

2. A small centrifugal pump is to run at 1600 rev/min and is to deliver 20 litres/s of water against a monometric head of 6 m. The flow enters the pump with zero whirl and the radial velocity component (v_f) remains constant through the impeller at 4 m/s. When the flow passes from the

impeller into the volute, 50% of the kinetic energy of flow (measured at impeller outlet) is converted to static pressure.

The impeller outlet diameter = (1·75 × inlet diameter) and the width of the flow passage at inlet = (0·4 × inlet diameter). Neglecting the vane thickness calculate:

(*a*) impeller inlet diameter;
(*b*) impeller outlet diameter;
(*c*) width of flow passage at inlet;
(*d*) width of flow passage at outlet;
(*e*) the vane angles at inlet and outlet.

3. The total head H_m metres corresponding to a discharge Q m³/s through a pipeline is given by

$$H_m = 14 + 600\,000Q^2$$

Which of the two pumps, A or B, is the most appropriate for this system? The liquid is water.

Performance data for pump A:

Discharge (litres/s)	0	6·94	11·6	16·6	23	28·3
Head (m)	32·6	35·8	33·2	28·8	21·2	14·1
Efficiency (%)	0	54	68	73	68	57

Performance data for pump B:

Discharge (litres/s)	0	9·7	16·3	24·0	33·0	39·7
Head (m)	30·6	34·0	31·5	27·0	20·0	13·4
Efficiency (%)	0	46	62	75	70	58

4. An inward flow Francis turbine is to fulfill the following specification:

Runner speed = 200 rev/min; Power output = 2 MW
Overall efficiency = 80%; Hydraulic efficiency = 85%
Runner outlet diameter = (0·65 × runner inlet diameter);
Runner tangential velocity at intake = $3{\cdot}24\sqrt{H}$;
Radial component of velocity of liquid,

$$v_f = 0{\cdot}731\sqrt{H} = \text{constant through runner}$$

Head supplied to turbine, H = 25 m

The flow is to discharge from the runner with zero whirl. Calculate:

(*a*) the inlet and outlet diameters of the runner;
(*b*) the angle of the guide vane;
(*c*) the inlet and outlet angles for the runner;
(*d*) the discharge required.

5. An axial flow pump is to be designed for a duty of 80 litres/s discharge and an ideal head of 10 m when handling water. The rotational speed is to be 1450 rev/min and the hub radius r_1 is to be half the tip radius. The whirl velocity distribution downstream of the impeller is in the form of a free vortex having a velocity at radius r of $w = 3{\cdot}2r_1/r$. The axial velocity may be assumed to be uniform.

Calculate the diameters of the impeller tip and hub.

APPENDIX

ANSWERS TO QUESTIONS

Chapter 1

1. 2·44 kNm; 0%.
2. 33·4 kN; 4·29 kNm.
3. 4·77 MN; 5·625 m.
4. 1732 kN/m; 23·4 degrees to horizontal.
5. 146·6 kN; 342 kN.
6. 2·81 m; $\theta = \tan^{-1}(0{\cdot}0229e)$.
7. (*a*) (*i*) 131·7 kN; 19·6 kN.
   (*ii*) 100·0 kN; 4·9 kN.
   (*b*) It sinks.

Chapter 2

2. 0·418 m^3/s.
3. 0·0302 Nm.
4. 1·258 N.
5. 1·310 kW.
6. 306 kN/m^2; 2·55 kW.
7. 73·4 hours.
8. 0·034 m^3/s; 37·5 kN/m^2.
9. 20·8 m^3/s; 8·59 MW.

Chapter 3

1. 0·634 m^3/s.
2. 0·0343 m^3/s; 173 s.
3. 0·892 m.
4. $C_v = 0{\cdot}926$; $C_D = 0{\cdot}589$; $C_c = 0{\cdot}636$.
5. 0·657 m.
6. 10·9 kN.
7. 397 N.

Chapter 4

1. 4·65 MN/m^2.
2. 28·0 MN/m^2; 3·0 s.
3. 0·135 m^3/s.
4. 876 N.
5. 8·82; 0·95; 9·89 kW; 497 N.
6. 1·85 MN/m^2; 63·4 MN/m^2.
8. 46·5 mm.

Chapter 6

1. 26·8 km/h; 1:128·25.
2. 1:5·48; 1:4930; 1:27 000; 1:5·48.
3. 9·85 kW.

4. 1·938 rev/s; 807·5 Nm.
5. 175·8 rev/min; 5·63 m^3/s; 15·13 m; 1·07 kW.
7. $1{:}\sqrt{5}$.

Chapter 7

1. 0·104 m. Duplicate a short length of pipe or use a short length of larger bore.
2. 0·762 m^3/s.
3. 20·5 kW; 0·179 m; 40·4 W.
4. 27%.
5. 103·8; 0.36, 0·21 and 0·15 m^3/s.
6. 0·427 m.
7.

Pipe	*I* (m^3/h)	*II* (m^3/h)
AB	481·25	4·667
BC	81·25	0·667
CD	−218·75	−2·333
DE	281·25	2·667
EF	−18·75	−0·333
FA	−518·75	−5·333

9. 22·12 m; 1·54, 1·49 and 3·03 m^3/s.

Chapter 8

2. $y = \dfrac{100\theta}{2\pi}$.
3. $\psi = -x^2 + z^2$: $\phi = 2xz$.
4. $\psi = -\dfrac{q\theta}{2\pi D}$; $\phi = \dfrac{kH}{2\pi D} \ln r$.
5. $\psi = -kxy$.
6. 2, $\frac{2}{3}$; 0·112 m.
7. 0·225 m.
8. $\psi = \dfrac{qy}{D} - \dfrac{Q\theta}{2\pi D}$; $\phi = \dfrac{qx}{D} - \dfrac{Q}{2\pi D} \ln r$.
9. $\psi = x + 20yz$ where $y =$ constant; $\psi = -(x^2 - y^2)$.
10. $\phi = ux + vy$.

Chapter 9

1. 178 613 J; 112 343 J.
3. 0·74 kg/s.
4. 350·9 m/s; 2·5%.
5. 1·01 kg/s; 572·1 kN/m^2.
6. (*a*) 0·009; (*b*) 100 kW; (*c*) 0·53.

Chapter 10

1. *See* p. 157.
2. $0{\cdot}273\rho U^2\delta$.

3. $\delta_1 = 0{\cdot}403x_1(R_e)^{-1/5}$.
4. 616 N.
5. *See* p. 165.
6. $\tau = \rho gmi$.
7. *See* p. 154; 57 mm.
8. 1.47 N.

Chapter 11

1. $1{:}\sqrt{5}$.
2. $0{\cdot}002\dot{7}$ m/s; $0{\cdot}02\dot{7}$.
3. *See* Example 50.
4. 0·009; 2·4%.
6. 0·635 m; 1370 m.
8. *See* p. 177.

Chapter 12

1. 21·5°; 55·4 Nm; 223·3 m; 286 mm.
2. (*a*) 63 mm; (*b*) 110·3 mm; (*c*) 25 mm; (*d*) 14·5 mm; (*e*) 37·1 and 76·4°.
3. Pump A.
4. (*a*) 1·547 m, 1·006 m; (*b*) 12·7°; (*c*) 46·2 and 19·1°; (*d*) 10·2 m^3/s.
5. 808 mm, 404 mm.

INDEX

Music
Tely
Computer end of November

450
800
1250
- 500
750
- 350
400
- 100
300 car - 200

700
550
200
100

400
700
400
1500
1000
1500
200

100

clothes
1000

£50
shirts 2
trousers 2
shoes 2